Patrick Moore's Practical Astronomy Series

DATE DUE

Other Titles in This Series

(*continued at the end of the book*)

Guide to Observing Deep-Sky Objects

A Complete Global Resource for Astronomers

Jeff A. Farinacci

 Includes CD-ROM

 Springer

Jeff A. Farinacci
DSObook@yahoo.com

Patrick Moore's Practical Astronomy Series ISSN 1617-7185

ISBN 978-0-387-72850-6   e-ISBN 978-0-387-72851-3

Library of Congress Control Number: 2007934533

The star charts in this book were created by Chris Marriott's SkyMap Pro. For more information, see
http://www.skymap.com

Printed on acid-free paper

9 8 7 6 5 4 3 2 1

springer.com

To my first teachers: my parents.
They taught me to read,
encouraged me at critical moments, and
gave me the opportunity to develop my abilities.
I shall be eternally indebted to them.

Contents

Contents

Introduction

When is the best time to see the Andromeda Galaxy, known also as M31? What about M42, the Great Nebula in Orion? That depends on several factors. First, where you are on the planet Earth, second, what time of year it is, and third, when you like to do your observing. The best time to see anything through our atmosphere is when the sky object is near the meridian, the imaginary line that runs through the North Pole point in the sky, through the zenith (the point directly overhead), and through the South Pole point in the sky (which will be below the horizon for Northern Hemispherians). Let's examine these three factors individually.

Since the Earth is roughly spherical, and since we have divided it up into roughly 24 time zones to match the 24 hour clock we use, where you are on the Earth is important. You will see a different sky if you were standing on the North Pole than if you were standing on the South Pole. Also, since the Earth rotates in a continuous fashion and the time zones are quantized, this also creates a problem. For example, say you were twenty feet east of the imaginary line which divides two time zones, and you had a partner with whom you were in contact that was forty feet west of you, so that they were twenty feet west of the time zone line. Hence, their clock would measure one hour earlier than yours, so that when a certain star crossed the meridian, they would record it an hour earlier than you did, even though the two of you saw this passage at virtually the same instant. Likewise, if you were on the far eastern edge of a time zone and you had a partner on the far western edge of the same time zone, you would record the meridian passage of a star nearly an hour earlier than your partner. So, it matters where on the Earth you stand.

Next, the time of year matters since in summer, the night-time side of the Earth faces 180° opposite of the direction the night-time side faces six months later, so that different stars are visible at a certain time of night. That means that the stars you could see at midnight in July are close to the Sun in January, and thus, can't be seen.

Finally, when you like to observe the Universe is also important. Do you prefer to drag the scope out just before bedtime, or are you an early bird, and you like to get up a few hours before dawn? Stars that are coming up in the East just after sunset are setting in the West in the hours before dawn. So, since the meridian provides the best spot to see things, when the sky object crosses this line is the best time. If you are a night owl, you will get the pleasure of seeing the object a few months later than if you were an early bird. Truly, the early bird does catch the nebula.

So, how does one use the charts in this book? For all the charts, the horizontal center of the graph represents midnight (more or less, again depending upon where you are in the time zone). The top of the graph and the bottom of the graph both represent 12:00 Noon. The horizontal axis represents the days throughout the year. You will see a shaded area across the bottom that bulges upwards in the middle of the year. This represents afternoon daylight with the top of the curve representing sunset. Mirror-image to this across the top is another shaded area that bulges downward. This represents morning daylight with the bottom curve representing dawn. Note that no correction is made for Daylight Saving Hours, only standard hours. Look at the first chart for Andromeda. For this chart and all charts, you will see lines that slope downward from left to right. For most constellations, these lines represent the times when the constellation is rising, crossing the meridian, or setting. How does one determine these times for constellations which represent extended (i.e., not point) objects? The author took an approximate central point to the constellation and used this point to determine the times. Looking at Andromeda, you see that it crosses the meridian shortly after sunset time in early to mid January, and finally sets at roughly 2 a.m. So if you wanted to see Andromeda at 4 a.m., you would be out of luck. But if you wanted to see Andromeda at 9 p.m., you could see it heading down in the West. Let's say that you are a night owl and want to see Andromeda between the hours of 10 p.m. and midnight, close to the meridian. October and November are your best months to see it. If you were an early bird and wanted to see it between the hours of 3 a.m. and 5 a.m., then mid-July to mid-September is your best bet, and since this is after the "rising" line, this will be on the Eastern side of the sky. You will also notice that for Andromeda, there is a grayed-out section sloping down from January to the middle of August. These sloping grayed-out areas are after the constellation has set and before it has risen, so it can't be seen. In general, any grayed-out section is a time when the constellation can't be seen, either due to daylight or being hidden by Earth. Now, some constellations are circumpolar (such as Ursa Minor, if you're far enough North). In these cases, there are no "rising" or "setting" lines, just the "meridian" line.

For the constellation drawings in the book, North is approximately up on the page. (I say approximately because for constellations that take up a large portion of the celestial sphere, North for the East end is at a different angle than for the West end, since the drawings are projections on a flat page and come from a sphere). The only three exceptions are Eridanus, Orion and Scorpius, where North points to the left side of the page. Included with each constellation rendering is a table of the stars and their positions and magnitudes, as well as a table of Deep sky objects, (all 110 Messier objects, 453 NGC, 1 Collinder cluster, 1 Melotte Cluster and the Large Magellanic Cloud, for a total of 566) brighter than tenth magnitude, as well as their sky positions and magnitudes.

About the Software

The CD-ROM accompanying this book has software written by the author. This software allows the user to determine what Deep Sky Objects (DSOs) are above the horizon for a time determined by the user. This CD-ROM contains the following files:

1. **deepsky.exe** This is the executable program. deepsky.exe requires the following three files to work.
2. **DeepSkyObjectList.txt** This is a text file that contains the database of the DSOs used by deepsky.exe.
3. **location.dat** This is a text file that contains the user's geographic location. This file is described in detail below.
4. **deltaT.dat** This is a text file that contains the difference between the Earth's Dynamical Time and Universal Time. This file is also described in detail below.

When the program deepsky.exe is run, it produces a file named DSO.dat, which is a text file that tells the user what DSOs are above the horizon for the location described in location.dat at the time input to the program. This file tells the user the name of the DSO, what constellation it is in, what kind of DSO it is (galaxy, star cluster, nebula), its sky location (Right ascension and Declination, as well as the Azimuth and Elevation), the visual magnitude, the angular diameter in arc-minutes, and a brief textual description of the DSO. Such a file allows a user to quickly and easily plan a night's observing session, whether it be after sunset, during the middle of the night, or before sunrise. The Azimuth in DSO. dat is to be regarded as a compass direction, measured in degrees. Azimuth = 0 is North, 90 is East, 180 is South, 270 is West. Elevation is the angular distance above the horizon, also measured in degrees. Thus, Elevation will range between 0 (right on the horizon) to 90 (being at the zenith point, directly overhead). If a DSO has an Azimuth of 105 degrees and an Elevation of 20 degrees, then the sky location of the DSO (at that given moment) is 20 degrees above the horizon spot that is 15 degrees South of East.

Description of the Files

The file location.dat contains the geographic information of the user's viewing location. This file can contain the geolocation information for many different spots, but only the first spot is used by the deepsky.exe program. If the user has several different spots from which they may do some observing, then the data may be kept in the file (recommendation: leave several blank lines between the blocks of data so as to easily see them) and, if another location is desired to be used for another run of deepsky.exe, then edit the location.dat file and move the desired location to the top of the file. The geolocation data in the file is written in five lines.

The first line contains the location's latitude, and has three numbers. The first number on the line is the latitude degrees. For Northern Hemisphere locations, the latitude degrees is positive. For Southern Hemisphere locations, the latitude degrees is negative. The second number on the line is the latitude minutes. This number should always be positive, even if the location is in the Southern Hemisphere. The third and final number on the first line is the latitude seconds. This number is also always positive. If using a Southern Hemisphere location, the program deepsky.exe makes the correct calculation for the geolocation by determining the sign of the latitude degrees (the first number).

The second line contains the location's longitude, and has three numbers. The first number on the line is the longitude degrees. For locations West of the prime meridian, the longitude degrees is negative. For locations East of the prime meridian, the longitude degrees is positive. The second number on the line is the longitude minutes. This number should always be positive, regardless of being East or West. The third and final number on the second line is the longitude seconds. This number is also always positive. If using a location West of the prime meridian (thus having a negative longitude), the program deepsky.exe makes the correct calculation for the geolocation by determining the sign of the longitude degrees (the first number).

The third line contains a single number, which is the elevation above sea level of the geolocation, in meters.

The fourth line contains a single number, which is the correction from Universal Time to Local Time. If the user needs to add 5 hours to Local Time in order to obtain Universal Time, then the number on the fourth line should be 5. If the user needs to subtract 2 hours from Local Time to obtain Universal Time, then the number on the fourth line should be -2. For those areas observing Daylight Saving Time, this number will need to be changed when the transition occurs from Standard Time to DST or back.

The fifth line is a text line describing the location. Blank spaces are allowed. Some examples might be

My Backyard
or
My Favorite Spot in the Mountains
or
The Astronomy Club's countryside viewing location.

This text line should be 70 characters or less in length. This text line is printed in the header of the output file DSO.dat, thus allowing the user to quickly see where the data is to be used from.

Here is an example location.dat file, showing two different locations. These blocks of data for these locations are separated by three blank lines. Comments are located following "<<<" on the lines, but should not be found in the file.

```
33 23 52.0           <<< Latitude 33 deg 23 min 52.0 sec North
-111 47 0.0          <<< Longitude 111 deg 47 min 0.0 sec West
378.0                <<< Elevation 378.0 meters above sea level
7                    <<< Add 7 hours to Local Time to get UT
```

Silvergate Park near home	<<< Text description of the location
	<<< blank line
	<<< blank line
	<<< blank line
37 6 55.0	<<< Latitude 37 deg 6 min 55.0 sec North
-94 29 0.0	<<< Longitude 94 deg 29 min 0.0 sec West
299.0	<<< Elevation 299.0 meters above sea level
5	<<< Add 5 hours to Local Time to get UT
Dad's House	<<< Text description of the location

Now, if the user wanted to switch locations, then edit this file and move the five lines for the desired location to the top of the file and save the file on the computer. When deepsky.exe is run again, it will use this location. So, edit this file and create your own library of locations for use in the deepsky program.

The file named deltaT.dat contains a single number, which is the difference between the Earth's Dynamical Time and Universal Time. This difference is in seconds. Since the Earth's rotation is not perfectly consistent, (Earth's rotation is slowing down, albeit at a very slow rate), there is a difference between the "perfect" time and the "actual" time as determined by the Earth's rotation. This difference is represented by the number in the deltaT.dat file. At the time of publication, this difference was 65.184 seconds. This is the constant 32.184 plus the current number of leap seconds (33 seconds as of January 1, 2006). This data was obtained from the website http://maia.usno.navy.mil/bulletin-a.html which is from the United States Naval Observatory. For future updates, look for the latest issue of IERS Bulletin A. If this number changes, then edit the file and put in the new number.

Running the Software

First, copy the four files described above into a directory (or folder) on your computer. This is how to create a new folder on your PC Desktop: right-click the mouse on the PC Desktop. This brings up a menu. Drag the mouse pointer down to "New", which brings up another menu. Slide the mouse over to "Folder" and click on this. This will create a new folder icon, with the name "New Folder" highlighted. You can type in a new name, such as "Astronomy", at this point to give your new folder a new name. When you have finished typing in a new name, you may either press the "Enter" key on the keyboard, or left-click the mouse to finish. Double-click the folder to open this new folder. Put the CD into your CD-ROM drive in your computer. Open the CD by double clicking on the icon labeled "My Computer" and then clicking on the icon for the CD Drive. You may then copy the files to the new folder you created by dragging the files to this new folder. Click on the icon for the program deepsky.exe. A window will open. See Figure 1.

At the top of the window is the geolocation information specified in the first five lines of the file location.dat. A description of the file DeepSkyObjectList.txt is printed next, telling the user how many Deep sky objects are in the list, along with how many are of each type. The user is prompted to input a Month, Day, Year, Hour and Minute for the computation. Use 1 for January, 2 for February, etc., for the month number and use a four-digit number for the year. For the hour, use

Figure 1. Running the deepsky.exe program

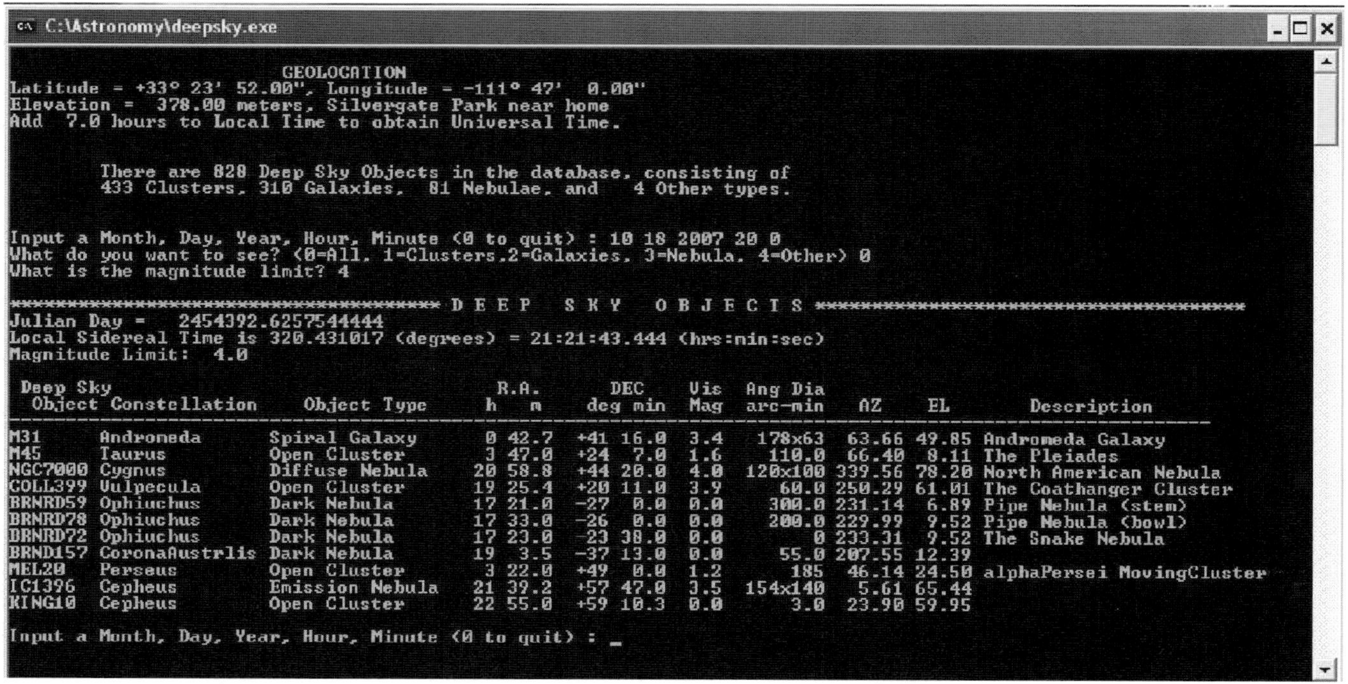

Figure 2. Results of running the deepsky.exe program

a integer from the 24 hour day, meaning, use 20 for 8 o'clock p.m. Once these 5 data are entered, press the Enter key. If you want to quit, simply enter 0 and press Enter. For example, to enter the time of October 18, 2007 at 8:00 p.m., enter 10 18 2007 20 0 <Enter>.

The user is next asked "What do you want to see?" Enter 0 to see a listing for all types, 1 for star clusters only, 2 for galaxies only, 3 for nebulae only and 4 for the other types only.

The user is then asked "What is the magnitude limit?" Only DSOs brighter than the magnitude limit that is input will be listed in the output file DSO.dat. When the Enter key is pressed, a listing of the DSOs visible at the geolocation and at the time entered will be produced in the window, as well as to the output file DSO.dat. When the user is done, enter 0 at the first prompt to end the deepsky.exe program. Then check the computer's directory for the file DSO.dat to see the output. The screen for a sample run with the magnitude limit of 4 is found in Figure 2.

A Note on the Deep Sky Object List file

Although many hours of research produced the database in the Deep sky object List, there are 13 DSOs for which a magnitude could not be determined. For these cases, a magnitude of 99.9 was put into the database. Furthermore, there are 828 DSOs in the database, which is 262 more than in the book. This database is in the file DeepSkyObjectList.txt.

How to read the Output file

When deepsky.exe is run, it produces an output file named DSO.dat in the same directory where the software is located. If you click on the icon for DSO.dat, this file should open in Microsoft Wordpad, although it can also be opened in Microsoft Word. If using Microsoft Word, you might have to set the page parameters, to view the file in its full width. This file contains the results of the calculations performed by deepsky. The first four lines are the geolocation header, which tells the user which location was used by deepsky. This will be the first geolocation in the file location.dat. Following this, a row of asterisks with DEEP SKY OBJECTS is printed out. The next line indicates the time requested for the output. The next line gives the local sidereal time (that is, the Right ascension) that is on the meridian at that moment. The next line lists the magnitude limit requested for the output. The DSO table header are on the next three lines. Then, a listing of all the DSOs that are above the horizon at that moment for that location. The DSO table consists of ten columns. The first column is the astronomical designation (such as M1 or NGC2237) of the DSO. The second column is the constellation in which the DSO appears. The third column tells the type of DSO, such as Globular Cluster, Open Cluster, Spiral Galaxy, etc. The fourth column is the Right ascension (R.A.) and has two numbers: the first are the hours (h) and the second are the minutes (m). The fifth column is the Declination (Dec) and has two numbers: the first are the degrees (deg) and the second are the minutes (min). The sixth column is the Visual Magnitude (Vis Mag) of the DSO. The seventh column is the Angular Diameter (Ang Dia) in arc-minutes (arc-min) for the DSO. The eighth column is the Azimuth (AZ) of the DSO at that moment. Azimuth is the "compass direction." North is Azimuth = 0, East is Azimuth = 90, South is Azimuth = 180, West is Azimuth = 270. Azimuth ranges from 0 (including 0) up to (but less than) 360. Thus, if the table has AZ = 75.00, then this is 15 degrees North of East. If the Azimuth is 210.50, then this is 30.5 degrees West of South, or 59.5 degrees South of West. The ninth column is the Elevation above the horizon (EL). If the Elevation is 90.00, then the DSO is directly overhead. If the Elevation is 30.00, then the DSO is 30 degrees above the horizon. The tenth column is a description of the DSO. Not all DSOs in the DeepSkyObjectList. txt file have descriptions for them. Some do, such as the Andromeda Galaxy or the Great Orion Nebula. The printout is 111 columns wide and should fit onto a standard sheet of paper, printed in Landscape Mode.

The Constellations

Andromeda

Princess of Ethiopia

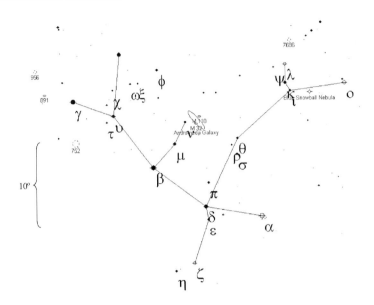

Star	Right ascension	Declination	Mag.
α	00h 08m 23.2s	+29° 05′ 26″	2.06
β	01h 09m 43.9s	+35° 37′ 14″	2.06
γ₁	02h 03m 53.9s	+42° 19′ 47″	2.26
γ₂	02h 03m 54.7s	+42° 19′ 51″	4.84
δ	00h 39m 19.6s	+30° 51′ 40″	3.27
ε	00h 38m 33.3s	+29° 18′ 43″	4.37
ζ	00h 47m 20.3s	+24° 16′ 02″	4.06
η	00h 57m 12.4s	+23° 25′ 04″	4.42
θ	00h 17m 05.4s	+38° 40′ 54″	4.61
ι	23h 38m 08.1s	+43° 16′ 05″	4.29
κ	23h 40m 24.4s	+44° 20′ 02″	4.14
λ	23h 37m 33.8s	+46° 27′ 30″	3.82
μ	00h 56m 45.1s	+38° 29′ 58″	3.87
ν	00h 49m 48.8s	+41° 04′ 44″	4.53
ξ	01h 22m 20.3s	+45° 31′ 44″	4.88
ο	23h 01m 55.2s	+42° 19′ 34″	3.62
π	00h 36m 52.8s	+33° 43′ 10″	4.36
ρ	00h 21m 07.2s	+37° 58′ 07″	5.18
σ	00h 18m 19.6s	+36° 47′ 07″	4.52
τ	01h 40m 34.7s	+40° 34′ 37″	4.94
υ	01h 36m 47.8s	+41° 24′ 20″	4.09
φ	01h 09m 30.1s	+47° 14′ 31″	4.25
χ	01h 39m 20.9s	+44° 23′ 10″	4.98
ψ	23h 46m 02.0s	+46° 25′ 13″	4.95
ω	01h 27m 39.2s	+45° 24′ 25″	4.83

Deep sky object	Description	Right ascension	Declination	Mag.
M31	Andromeda Galaxy	00h 42m 42s	+41° 16.0′	3.4
M32	Satellite to M31	00h 42m 42s	+40° 52.0′	8.1
M110	Satellite to M31	00h 40m 24s	+41° 41.0′	8.5
NGC 752	Open Cluster	01h 57m 48s	+37° 41.0′	5.7
NGC 891	Edge-On Spiral Galaxy	02h 22m 33s	+42° 21.0′	10.0
NGC 956	Open Cluster	02h 32m 31s	+44° 35.6′	8.9
NGC 7662	The Blue Snowball Nebula	23h 25m 54s	+42° 33.0′	9.0
NGC 7686	Open Cluster	23h 30m 12s	+49° 08.0′	5.6

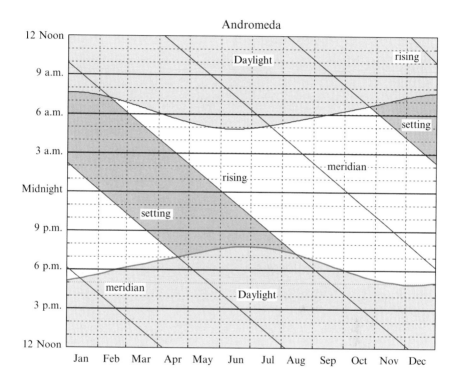

Antlia

the Air Pump

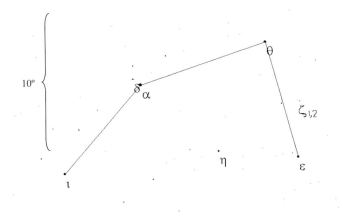

Star	Right ascension	Declination	Mag.
α	10h 27m 09.1s	$-31°$ 04′ 04″	4.25
δ	10h 29m 35.3s	$-30°$ 36′ 26″	5.56
ε	09h 29m 14.7s	$-35°$ 57′ 06″	4.51
ζ_1	09h 30m 45.4s	$-31°$ 53′ 29″	7.21
ζ_2	09h 30m 46.1s	$-31°$ 53′ 22″	6.35
ζ_3	09h 31m 32.2s	$-31°$ 52′ 18″	5.93
η	09h 58m 52.2s	$-35°$ 53′ 28″	5.23
θ	09h 44m 12.1s	$-27°$ 46′ 10″	4.79
ι	10h 56m 43.0s	$-37°$ 08′ 16″	4.60

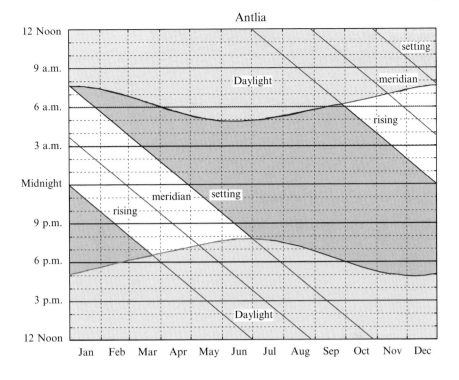

Antlia

Apus
the Bird of Paradise

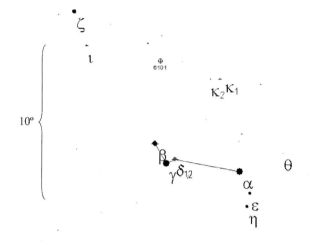

Star	Right ascension	Declination	Mag.
α	14h 47m 51.6s	−79° 02′ 41″	3.83
β	16h 43m 04.4s	−77° 31′ 03″	4.24
γ	16h 33m 27.1s	−78° 53′ 49″	3.89
δ_1	16h 20m 20.7s	−78° 41′ 45″	4.68
δ_2	16h 20m 26.7s	−78° 40′ 02″	5.27
ε	14h 22m 22.7s	−80° 06′ 32″	5.06
ζ	17h 21m 59.5s	−67° 46′ 13″	4.78
η	14h 18m 13.6s	−81° 00′ 27″	4.91
θ	14h 05m 19.8s	−76° 47′ 48″	5.50
ι	17h 22m 05.8s	−70° 07′ 24″	5.41
κ_1	15h 31m 30.8s	−73° 23′ 22″	5.49
κ_2	15h 40m 21.2s	−73° 26′ 48″	5.65

Deep sky object	Description	Right ascension	Declination	Mag.
NGC 6101	Globular Cluster	16h 25m 48s	−72° 12.1′	9.3

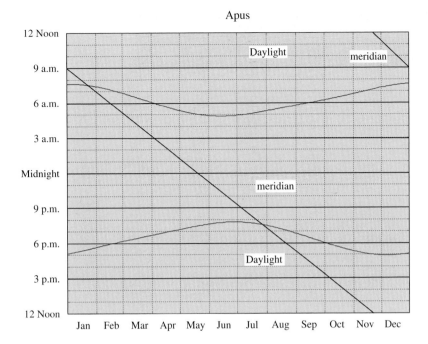

Apus

The constellation Apus is not visible from mid-northern latitudes

Aquarius

the Water Bearer

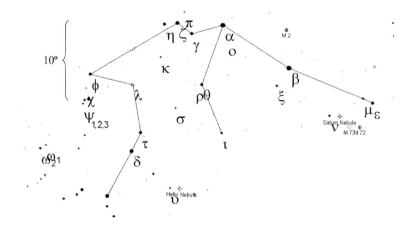

Star	Right ascension	Declination	Mag.
α	22h 05m 46.9s	−00° 19′ 11″	2.96
β	21h 31m 33.4s	−05° 34′ 16″	2.91
γ	22h 21m 39.3s	−01° 23′ 14″	3.84
δ	22h 54m 38.9s	−15° 49′ 15″	3.27
ε	20h 47m 40.5s	−09° 29′ 45″	3.77
ζ_1	22h 28m 50.0s	−00° 01′ 12″	4.42
ζ_2	22h 28m 49.6s	−00° 01′ 13″	4.59
η	22h 35m 21.3s	−00° 07′ 03″	4.02
θ	22h 16m 49.9s	−07° 47′ 00″	4.16
ι	22h 06m 26.1s	−13° 52′ 11″	4.27
κ	22h 37m 45.3s	−04° 13′ 41″	5.03
λ	22h 52m 36.8s	−07° 34′ 47″	3.74
μ	20h 52m 39.1s	−08° 59′ 00″	4.73
ν	21h 09m 35.5s	−11° 22′ 18″	4.51
ξ	21h 37m 45.0s	−07° 51′ 15″	4.69
o	22h 03m 18.7s	−02° 09′ 19″	4.69
π	22h 25m 16.5s	+01° 22′ 39″	4.66
ρ	22h 20m 11.8s	−07° 49′ 16″	5.37
σ	22h 30m 38.7s	−10° 40′ 41″	4.82
τ_1	22h 49m 35.4s	−13° 35′ 33″	4.01
τ_2	22h 47m 42.7s	−14° 03′ 23″	5.66
υ	22h 34m 41.6s	−20° 42′ 30″	5.20
φ	23h 14m 19.3s	−06° 02′ 56″	4.22
χ	23h 16m 50.8s	−07° 43′ 36″	5.06
ψ_1	23h 15m 53.4s	−09° 05′ 16″	4.21
ψ_2	23h 17m 54.1s	−09° 10′ 57″	4.39
ψ_3	23h 18m 57.6s	−09° 36′ 38″	4.98
ω_1	23h 42m 43.2s	−14° 32′ 42″	4.49
ω_2	23h 39m 47.0s	−14° 13′ 18″	5.00

Deep sky object	Description	Right ascension	Declination	Mag.
M2	Globular Cluster	21h 33m 29.3s	–00° 49.4′	6.4
M72	Globular Cluster	20h 53m 30.0s	–12° 32.0′	9.3
M73	Open Cluster	20h 59m 00.0s	–12° 38.0′	8.9
NGC 7009	The Saturn Nebula	21h 04m 10.7s	–11° 21.6′	8.0
NGC 7293	The Helix Nebula	22h 29m 38.4s	–20° 50.2′	7.3

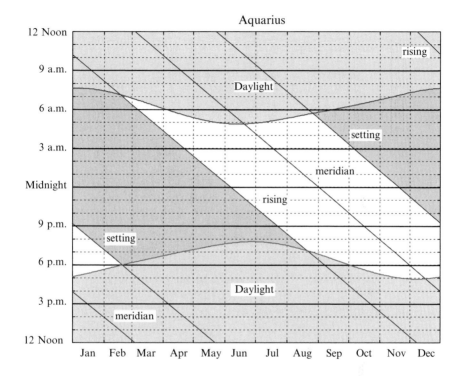

Aquila

the Eagle

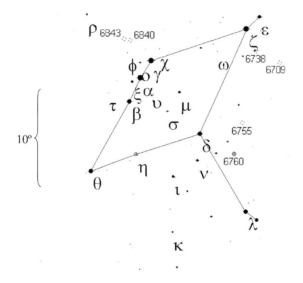

Star	Right ascension	Declination	Mag.
α	19h 50m 46.9s	+08° 52′ 06″	0.77
β	19h 55m 18.7s	+06° 24′ 24″	3.71
γ	19h 46m 15.5s	+10° 36′ 48″	2.72
δ	19h 25m 29.8s	+03° 06′ 53″	3.36
ε	18h 59m 37.3s	+15° 04′ 06″	4.02
ζ	19h 05m 24.5s	+13° 51′ 48″	2.99
η	19h 52m 28.3s	+01° 00′ 20″	3.90
θ	20h 11m 18.2s	−00° 49′ 17″	3.23
ι	19h 36m 43.2s	−01° 17′ 11″	4.36
κ	19h 36m 53.4s	−07° 01′ 39″	4.95
λ	19h 06m 14.8s	−04° 52′ 57″	3.44
μ	19h 34m 05.3s	+07° 22′ 44″	4.45
ν	19h 26m 31.0s	+00° 20′ 19″	4.66
ξ	19h 54m 14.8s	+08° 27′ 41″	4.71
ο	19h 51m 01.5s	+10° 24′ 56″	5.11
π	19h 48m 42.0s	+11° 48′ 57″	5.72
ρ	20h 14m 16.5s	+15° 11′ 51″	4.95
σ	19h 39m 11.5s	+05° 23′ 52″	5.17
τ	20h 04m 08.2s	+07° 16′ 41″	5.52
υ	19h 45m 39.8s	+07° 36′ 48″	5.91
φ	19h 56m 14.1s	+11° 25′ 25″	5.28
χ	19h 42m 33.9s	+11° 49′ 36″	5.27
ψ	19h 44m 34.0s	+13° 18′ 10″	6.26
ω₁	19h 17m 48.9s	+11° 35′ 44″	5.28
ω₂	19h 19m 52.9s	+11° 32′ 06″	6.02

Deep sky object	Description	Right ascension	Declination	Mag.
NGC 6709	Open Cluster	18h 51m 30.0s	+10° 21.0′	6.7
NGC 6738	Open Cluster	19h 01m 24.0s	+11° 36.0′	8.3
NGC 6755	Open Cluster	19h 07m 48.0s	+04° 14.0′	7.5
NGC 6760	Globular Cluster	19h 11m 12.0s	+01° 02.0′	9.1
NGC 6840	Open Cluster	19h 55m 18.0s	+12° 06.0′	10.0
NGC 6843	Open Cluster	19h 56m 06.0s	+12° 09.0′	10.0

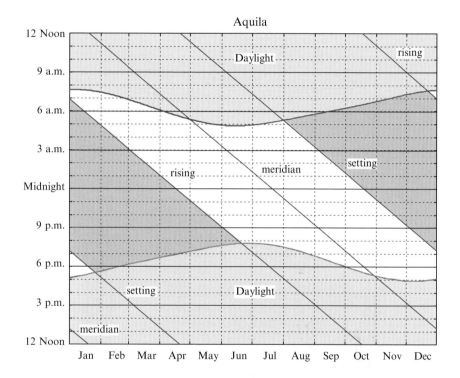

Aquila

Ara

the Altar

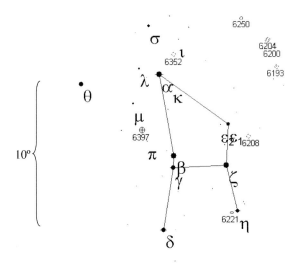

Star	Right ascension	Declination	Mag.
α	17h 31m 50.4s	−49° 52′ 34″	2.95
β	17h 25m 17.9s	−55° 31′ 47″	2.85
γ	17h 25m 23.5s	−56° 22′ 39″	3.34
δ	17h 31m 05.9s	−60° 41′ 01″	3.62
ε_1	16h 59m 35.0s	−53° 09′ 38″	4.06
ε_2	17h 03m 08.6s	−53° 14′ 13″	5.29
ζ	16h 58m 37.1s	−55° 59′ 24″	3.13
η	16h 49m 47.0s	−59° 02′ 29″	3.76
θ	18h 06m 37.7s	−50° 05′ 30″	3.66
ι	17h 23m 16.0s	−47° 28′ 05″	5.25
κ	17h 25m 59.9s	−50° 38′ 01″	5.23
λ	17h 40m 23.4s	−49° 24′ 56″	4.77
μ	17h 44m 08.6s	−51° 50′ 03″	5.15
π	17h 38m 05.5s	−54° 30′ 01″	5.25
σ	17h 35m 39.4s	−46° 30′ 20″	4.59

Deep sky object	Description	Right ascension	Declination	Mag.
NGC 6193	Open Cluster	16h 41m 20.2s	−48° 45.8′	5.2
NGC 6200	Open Cluster	16h 44m 12.0s	−47° 29.0′	7.4
NGC 6204	Open Cluster	16h 46m 30.0s	−47° 01.0′	8.2
NGC 6208	Open Cluster	16h 49m 30.0s	−53° 49.0′	7.2
NGC 6250	Open Cluster	16h 58m 00.0s	−45° 48.0′	5.9
NGC 6352	Globular Cluster	17h 25m 30.0s	−48° 25.0′	8.1
NGC 6362	Globular Cluster	17h 31m 54.0s	−67° 03.0′	8.3
NGC 6397	Globular Cluster	17h 40m 42.0s	−53° 40.0′	5.7

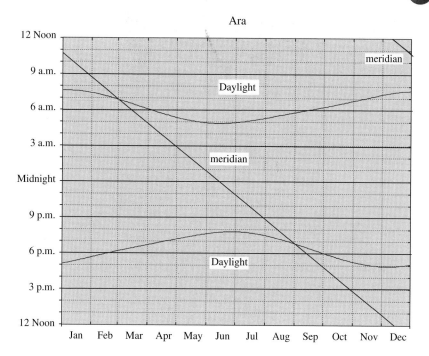

The constellation Ara is not visible from mid-northern latitudes

Aries

the Ram

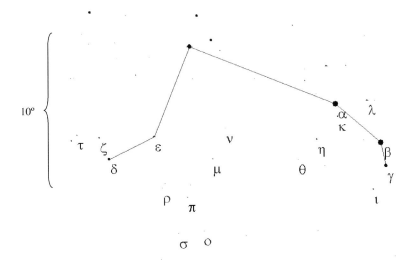

Star	Right ascension	Declination	Mag.
α	02h 07m 10.3s	+23° 27′ 45″	2.00
β	01h 54m 38.3s	+20° 48′ 29″	2.64
γ₁	01h 53m 31.8s	+19° 17′ 37″	4.75
γ₂	01h 53m 31.7s	+19° 17′ 45″	4.83
δ	03h 11m 37.7s	+19° 43′ 36″	4.35
ε₁	02h 59m 12.6s	+21° 20′ 25″	4.63
ε₂	02h 59m 12.6s	+21° 20′ 25″	4.63
ζ	03h 14m 54.0s	+21° 02′ 40″	4.89
η	02h 12m 48.0s	+21° 12′ 39″	5.27
θ	02h 18m 07.5s	+19° 54′ 04″	5.62
ι	01h 57m 21.0s	+17° 49′ 03″	5.10
κ	02h 06m 33.8s	+22° 38′ 54″	5.03
λ	01h 57m 55.7s	+23° 35′ 46″	4.79
μ	02h 42m 21.9s	+20° 00′ 42″	5.69
ν	02h 38m 48.9s	+21° 57′ 41″	5.30
ξ	02h 24m 49.0s	+10° 36′ 38″	5.47
o	02h 44m 32.9s	+15° 18′ 42″	5.77
π	02h 49m 17.5s	+17° 27′ 51″	5.22
ρ₁	02h 55m 48.4s	+18° 19′ 54″	5.91
ρ₂	02h 56m 26.1s	+18° 01′ 23″	5.63
σ	02h 51m 29.5s	+15° 04′ 55″	5.49
τ₁	03h 22m 45.2s	+20° 44′ 31″	5.09
τ₂	03h 21m 13.6s	+21° 08′ 49″	5.28

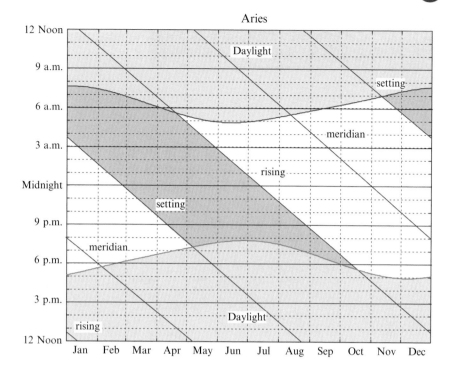

Auriga
the Charioteer

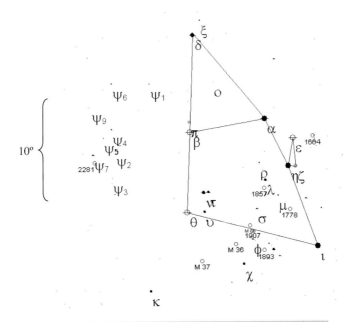

Star	Right ascension	Declination	Mag.
α	05h 16m 41.3s	+45° 59′ 53″	0.08
β	05h 59m 31.7s	+44° 56′ 51″	1.90
δ	05h 59m 31.6s	+54° 17′ 05″	3.72
ε	05h 01m 58.1s	+43° 49′ 24″	2.99
ζ	05h 02m 28.6s	+41° 04′ 33″	3.75
η	05h 06m 30.8s	+41° 14′ 04″	3.17
θ	05h 59m 43.2s	+37° 12′ 45″	2.62
ι	04h 56m 59.6s	+33° 09′ 58″	2.69
κ	06h 15m 22.6s	+29° 29′ 53″	4.35
λ	05h 19m 08.4s	+40° 05′ 57″	4.71
μ	05h 13m 25.6s	+38° 29′ 04″	4.86
ν	05h 51m 29.3s	+39° 08′ 55″	3.97
ξ	05h 54m 50.7s	+55° 42′ 25″	4.99
ο	05h 45m 54.0s	+49° 49′ 35″	5.47
π	05h 59m 56.1s	+45° 56′ 13″	4.26
ρ	05h 21m 48.4s	+41° 48′ 17″	5.23
σ	05h 24m 39.1s	+37° 23′ 08″	4.99
τ	05h 49m 10.4s	+39° 10′ 52″	4.52
υ	05h 51m 02.4s	+37° 18′ 20″	4.74
φ	05h 27m 38.8s	+34° 28′ 33″	5.07
χ	05h 32m 43.6s	+32° 11′ 31″	4.76
ψ₁	06h 39m 19.8s	+42° 29′ 20″	4.79
ψ₂	06h 24m 53.8s	+49° 17′ 17″	4.91
ψ₃	06h 38m 49.1s	+39° 54′ 09″	5.20
ψ₄	06h 43m 04.9s	+44° 31′ 28″	5.02
ψ₅	06h 46m 44.3s	+43° 34′ 39″	5.25

Star	Right ascension	Declination	Mag.
ψ₆	06h 47m 39.5s	+48° 47′ 22″	5.22
ψ₇	06h 50m 45.9s	+41° 46′ 53″	5.02
ψ₈	06h 53m 56.9s	+38° 30′ 18″	6.48
ψ₉	06h 56m 32.2s	+46° 16′ 27″	5.87
ω	04h 59m 15.3s	+37° 53′ 25″	4.94

Deep sky object	Description	Right ascension	Declination	Mag.
M36	Open Cluster	05h 36m 17.7s	+34° 8.4′	6.0
M37	Open Cluster	05h 52m 18.3s	+32° 33.2′	5.6
M38	Open Cluster	05h 28m 42.5s	+35° 51.3′	6.4
NGC 1664	Open Cluster	04h 51m 08.4s	+43° 40.5′	7.6
NGC 1778	Open Cluster	05h 08m 05.7s	+37° 01.4′	7.7
NGC 1857	Open Cluster	05h 20m 05.5s	+39° 20.6′	7.0
NGC 1893	Open Cluster	05h 22m 45.1s	+33° 25.2′	7.5
NGC 1907	Open Cluster	05h 28m 04.5s	+35° 19.5′	8.2
NGC 2281	Open Cluster	06h 48m 17.8s	+41° 04.7′	5.4

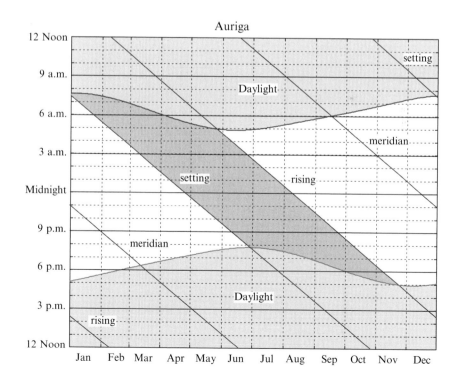

Boötes
the Bear Driver

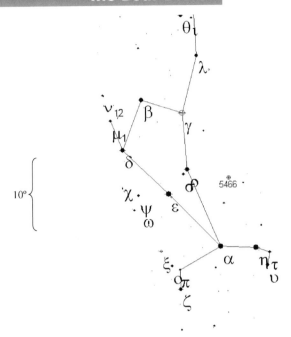

Star	Right ascension	Declination	Mag.
α	14h 15m 39.6s	+19° 10′ 57″	−0.04
β	15h 01m 56.6s	+40° 23′ 26″	3.50
γ	14h 32m 04.6s	+38° 18′ 29″	3.03
δ	15h 15m 30.1s	+33° 18′ 53″	3.47
ε_1	14h 44m 59.1s	+27° 04′ 27″	2.70
ε_2	14h 44m 59.1s	+27° 04′ 30″	5.12
ζ_1	14h 41m 08.8s	+13° 43′ 42″	4.43
ζ_2	14h 41m 08.8s	+13° 43′ 42″	4.83
η	13h 54m 41.0s	+18° 23′ 52″	2.68
θ	14h 25m 11.7s	+51° 51′ 03″	4.05
ι	14h 16m 09.8s	+51° 22′ 02″	4.75
κ_1	14h 13m 28.9s	+51° 47′ 24″	4.54
κ_2	14h 13m 27.6s	+51° 47′ 15″	6.69
λ	14h 16m 22.9s	+46° 05′ 18″	4.18
μ_1	15h 24m 29.3s	+37° 22′ 38″	4.31
μ_2	15h 24m 30.8s	+37° 20′ 51″	6.50
ν_1	15h 30m 55.7s	+40° 49′ 59″	5.02
ν_2	15h 31m 46.9s	+40° 53′ 58″	5.02
ξ	14h 51m 23.2s	+19° 06′ 04″	4.55
o	14h 45m 14.4s	+16° 57′ 52″	4.60
π_1	14h 40m 43.5s	+16° 25′ 06″	4.94
π_2	14h 40m 43.8s	+16° 25′ 04″	5.81

Star	Right ascension	Declination	Mag.
ρ	14h 31m 49.7s	+30° 22′ 17″	3.58
σ	14h 34m 40.7s	+29° 44′ 42″	4.46
τ	13h 47m 15.7s	+17° 27′ 24″	4.50
υ	13h 49m 28.5s	+15° 47′ 52″	4.07
φ	15h 37m 49.5s	+40° 21′ 12″	5.24
χ	15h 14m 29.1s	+29° 09′ 51″	5.26
ψ	15h 04m 26.7s	+26° 56′ 51″	4.54
ω	15h 02m 06.4s	+25° 00′ 29″	4.81

Deep sky object	Description	Right ascension	Declination	Mag.
NGC 5466	Globular Cluster	14h 05m 30.0s	+28° 32.0′	9.0

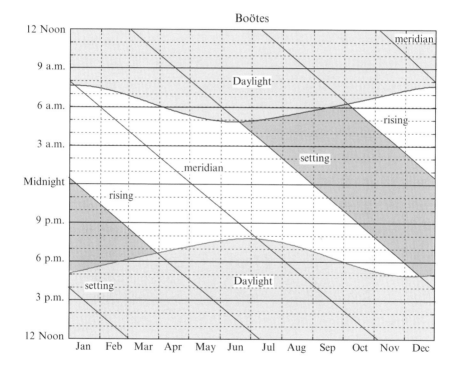

Boötes

Caelum
the Sculptor's Chisel

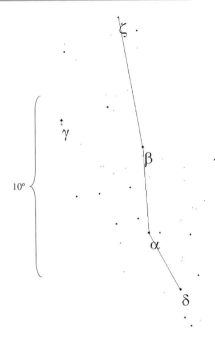

Star	Right ascension	Declination	Mag.
α	04h 40m 33.6s	−41° 51′ 50″	4.45
β	04h 42m 03.4s	−37° 08′ 40″	5.05
γ₁	05h 04m 24.3s	−35° 29′ 00″	4.55
γ₂	05h 04m 26.0s	−35° 42′ 19″	6.34
δ	04h 30m 50.1s	−44° 57′ 14″	5.07
ζ	04h 47m 49.6s	−30° 01′ 13″	6.37

Caelum

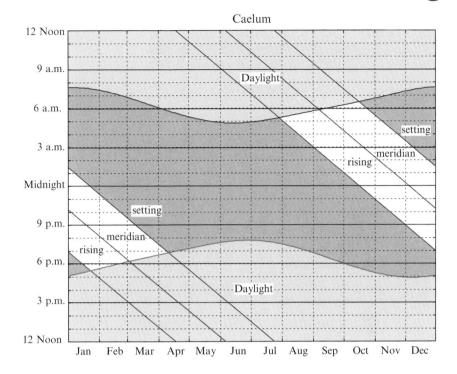

Camelopardalis
the Giraffe

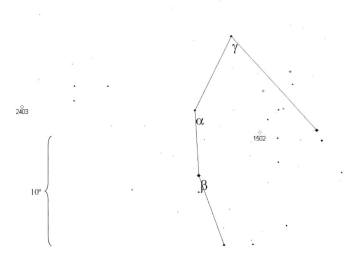

Star	Right ascension	Declination	Mag.
α	04h 54m 03.0s	+66° 20′ 34″	4.29
β	05h 03m 25.1s	+60° 26′ 32″	4.03
γ	03h 50m 21.5s	+71° 19′ 57″	4.63

Deep sky object	Description	Right ascension	Declination	Mag.
NGC 1502	Open Cluster	04h 07m 49.3s	+62° 19.9′	6.9
NGC 2403	Spiral Galaxy	07h 36m 51.8s	+65° 36.2′	8.4

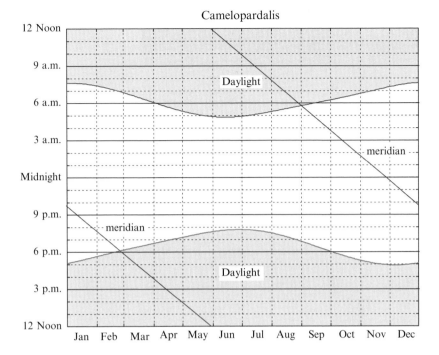

Camelopardalis

Cancer

the Crab

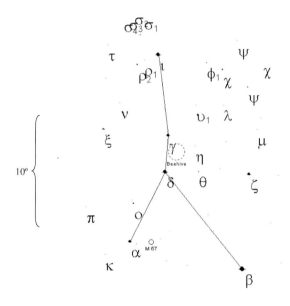

Star	Right ascension	Declination	Mag.
α	08h 58m 29.2s	+11° 51′ 28″	4.25
β	08h 16m 30.9s	+ 9° 11′ 08″	3.52
γ	08h 43m 17.1s	+21° 28′ 06″	4.66
δ	08h 44m 41.0s	+18° 09′ 15″	3.94
ε	08h 40m 26.9s	+19° 32′ 42″	6.30
ζ_1	08h 12m 12.6s	+17° 38′ 52″	5.44
ζ_2	08h 12m 12.6s	+17° 38′ 52″	6.20
ζ_3	08h 12m 13.2s	+17° 38′ 51″	6.01
η	08h 32m 42.4s	+20° 26′ 28″	5.33
θ	08h 31m 35.7s	+18° 05′ 40″	5.35
ι_1	08h 46m 41.8s	+28° 45′ 36″	4.02
ι_2	08h 46m 40.0s	+28° 45′ 55″	6.57
κ	09h 07m 44.8s	+10° 40′ 06″	5.24
λ	08h 20m 32.1s	+24° 01′ 20″	5.98
μ_1	08h 07m 45.8s	+21° 34′ 54″	5.30
μ_2	08h 06m 18.3s	+22° 38′ 08″	5.99
ν	09h 02m 44.2s	+24° 27′ 11″	5.45
ξ	09h 09m 21.5s	+22° 02′ 43″	5.14
o	08h 57m 14.9s	+15° 19′ 21″	5.20
o	08h 57m 35.1s	+15° 34′ 53″	5.67
π_1	09h 15m 13.8s	+14° 56′ 29″	5.34
π_2	09h 12m 17.5s	+14° 59′ 46″	6.51
ρ_1	08h 55m 39.6s	+27° 55′ 39″	5.22
ρ_2	08h 55m 39.6s	+27° 55′ 39″	5.22
σ_1	08h 59m 32.6s	+32° 25′ 07″	5.20

Star	Right ascension	Declination	Mag.
σ_2	08h 52m 34.6s	+32° 28′ 27″	5.66
σ_3	08h 56m 56.6s	+32° 54′ 37″	5.45
τ	09h 08m 00.0s	+29° 39′ 15″	5.43
υ_1	08h 31m 30.4s	+24° 04′ 52″	5.75
υ_2	08h 33m 00.0s	+24° 05′ 05″	6.36
φ_1	08h 26m 27.6s	+27° 53′ 36″	5.57
φ_2	08h 26m 46.7s	+26° 56′ 04″	6.32
φ_3	08h 26m 47.0s	+26° 56′ 07″	6.30
χ	08h 20m 03.8s	+27° 13′ 03″	5.14
ψ	08h 10m 27.1s	+25° 30′ 26″	5.73
ω_1	08h 00m 55.8s	+25° 23′ 34″	5.83
ω_2	08h 01m 43.7s	+25° 05′ 23″	6.31

Deep sky object	Description	Right ascension	Declination	Mag.
M44	Beehive Cluster	08h 40m 22.2s	+19° 40.3′	3.1
M67	Open Cluster	08h 51m 20.1s	+11° 48.7′	6.9

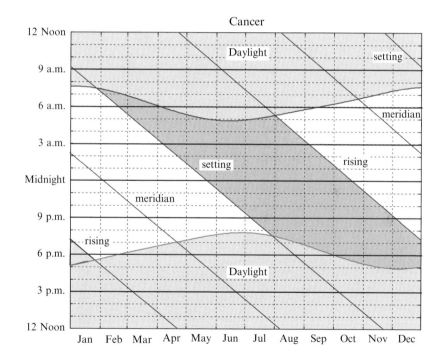

Canes Venatici
the Hunting Dogs

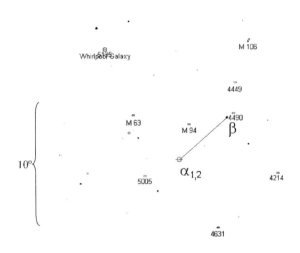

Star	Right ascension	Declination	Mag.
α_1	12h 56m 01.6s	+38° 19′ 06″	2.90
α_2	12h 56m 00.3s	+38° 18′ 53″	5.60
β	12h 33m 44.5s	+41° 21′ 27″	4.26

Deep sky object	Description	Right ascension	Declination	Mag.
M3	Globular Cluster	13h 42m 12.0s	+28° 23.0′	5.9
M51	Whirlpool Galaxy	13h 29m 52.1s	+47° 11.7′	8.4
M63	Sunflower Galaxy	13h 15m 49.1s	+42° 01.8′	8.6
M94	Spiral Galaxy	12h 50m 54.0s	+41° 07.0′	8.1
M106	Spiral Galaxy	12h 18m 57.5s	+47° 18.2′	8.4
NGC 4214	Galaxy	12h 15m 39.5s	+36° 19.6′	9.7
NGC 4449	Irregular Galaxy	12h 28m 12.0s	+44° 06.0′	9.4
NGC 4490	The Cocoon Galaxy	12h 30m 36.0s	+41° 38.0′	9.8
NGC 4631	The Whale Galaxy	12h 42m 06.0s	+32° 33.0′	9.8
NGC 5005	Galaxy	13h 10m 56.6s	+37° 03.5′	9.8
NGC 5195	Galaxy	13h 30m 00.0s	+47° 16.0′	9.6

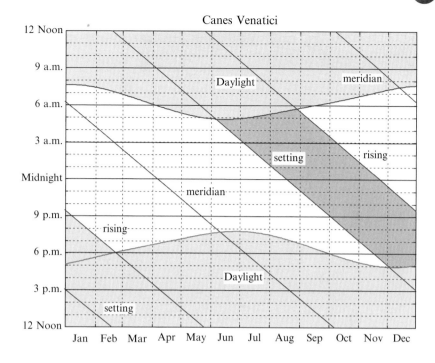

Canes Venatici

Canis Major
the Big Dog

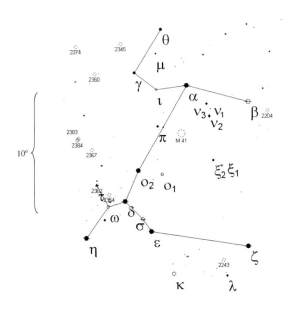

Star	Right ascension	Declination	Mag.
α	06h 45m 08.9s	−16° 42′ 58″	−1.46
β	06h 22m 41.9s	−17° 57′ 22″	1.98
γ	07h 03m 45.4s	−15° 38′ 00″	4.12
δ	07h 08m 23.4s	−26° 23′ 35″	1.84
ε	06h 58m 37.5s	−28° 58′ 20″	1.50
ζ	06h 20m 18.7s	−30° 03′ 48″	3.02
η	07h 24m 05.6s	−29° 18′ 11″	2.45
θ	06h 54m 11.3s	−12° 02′ 19″	4.07
ι	06h 56m 08.1s	−17° 03′ 15″	4.37
κ	06h 49m 50.4s	−32° 30′ 31″	3.96
λ	06h 28m 10.1s	−32° 34′ 49″	4.48
μ	06h 56m 06.6s	−14° 02′ 37″	5.00
ν_1	06h 36m 41.0s	−19° 15′ 22″	3.95
ν_2	06h 36m 22.8s	−18° 39′ 36″	5.70
ν_3	06h 37m 53.3s	−18° 14′ 15″	4.43
ξ_1	06h 31m 51.3s	−23° 25′ 06″	4.33
ξ_2	06h 35m 03.3s	−22° 57′ 53″	4.54
o_1	07h 03m 01.4s	−23° 50′ 00″	3.02
o_1	06h 54m 07.8s	−24° 11′ 02″	3.87
π	06h 55m 37.3s	−20° 08′ 11″	4.68
σ	07h 01m 43.1s	−27° 56′ 06″	3.47
τ	07h 18m 42.4s	−24° 57′ 15″	4.40
ω	07h 14m 48.6s	−26° 46′ 22″	3.85

Deep sky object	Description	Right ascension	Declination	Mag.
M41	Open Cluster	06h 46m 00.0s	−20° 45.3′	4.5
NGC 2204	Open Cluster	06h 15m 32.2s	−18° 39.9′	8.6
NGC 2243	Open Cluster	06h 29m 34.5s	−31° 16.9′	9.4
NGC 2345	Open Cluster	07h 08m 18.8s	−13° 11.6′	7.7
NGC 2354	Open Cluster	07h 14m 15.3s	−25° 41.5′	6.5
NGC 2360	Open Cluster	07h 17m 43.1s	−15° 38.5′	7.2
NGC 2362	Mexican Jumping Star Cluster	07h 18m 41.5s	−24° 57.3′	4.1
NGC 2367	Open Cluster	07h 20m 04.5s	−21° 53.1′	7.9
NGC 2374	Open Cluster	07h 23m 56.1s	−13° 15.8′	8.0
NGC 2383	Open Cluster	07h 24m 39.9s	−20° 56.8′	8.4
NGC 2384	Open Cluster	07h 25m 11.8s	−21° 01.4′	7.4

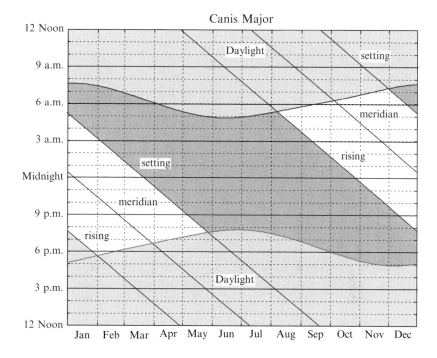

Canis Major

Canis Minor

the Little Dog

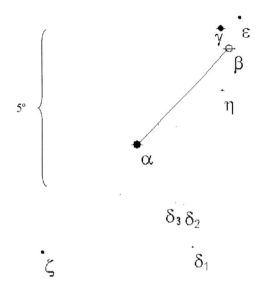

Star	Right ascension	Declination	Mag.
α	07h 39m 18.1s	+05° 13′ 30″	0.38
β	07h 27m 09.0s	+08° 17′ 21″	2.90
γ	07h 28m 09.7s	+08° 55′ 32″	4.32
δ₁	07h 32m 05.8s	+01° 54′ 52″	5.25
δ₂	07h 33m 11.6s	+03° 17′ 25″	5.59
δ₃	07h 34m 15.8s	+03° 22′ 17″	5.81
ε	07h 25m 38.8s	+09° 16′ 34″	4.99
ζ	07h 51m 41.9s	+01° 46′ 01″	5.14
η	07h 28m 02.0s	+06° 56′ 31″	5.25

Canis Minor

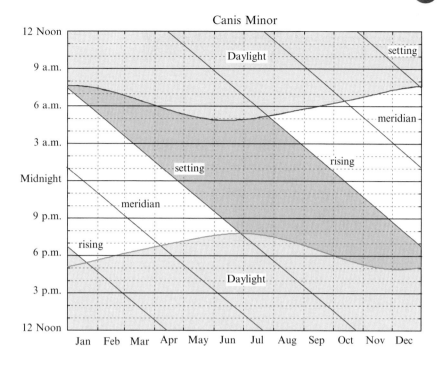

Capricornus

the Sea Goat

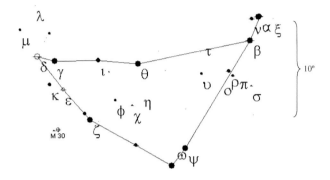

Star	Right ascension	Declination	Mag.
α_1	20h 18m 03.2s	−12° 32′ 42″	3.57
α_2	20h 17m 38.8s	−12° 30′ 30″	4.24
β	20h 21m 00.6s	−14° 46′ 53″	3.08
γ	21h 40m 05.4s	−16° 39′ 45″	3.68
δ	21h 47m 02.3s	−16° 07′ 38″	2.87
ε	21h 37m 04.7s	−19° 27′ 58″	4.68
ζ	21h 26m 39.9s	−22° 24′ 41″	3.74
η	21h 04m 24.2s	−19° 51′ 18″	4.84
θ	21h 05m 56.7s	−17° 13′ 58″	4.07
ι	21h 22m 14.7s	−16° 50′ 05″	4.28
κ	21h 42m 39.4s	−18° 51′ 59″	4.73
λ	21h 46m 32.0s	−11° 21′ 58″	5.58
μ	21h 53m 17.7s	−13° 33′ 07″	5.08
ν	20h 20m 39.7s	−12° 45′ 33″	4.76
ξ_1	20h 12m 25.8s	−12° 37′ 03″	5.85
ξ_2	20h 11m 57.8s	−12° 23′ 33″	6.34
o_1	20h 29m 53.8s	−18° 35′ 00″	5.94
o_2	20h 29m 52.4s	−18° 35′ 12″	6.74
π	20h 27m 19.1s	−18° 12′ 42″	5.25
ρ	20h 28m 51.5s	−17° 48′ 49″	4.78
σ	20h 19m 23.5s	−19° 07′ 07″	5.28
τ	20h 39m 16.3s	−14° 57′ 17″	5.22
υ	20h 40m 02.9s	−18° 08′ 19″	5.10
φ	21h 15m 37.8s	−20° 39′ 06″	5.24
χ	21h 08m 33.5s	−21° 11′ 37″	6.02
ψ	20h 46m 05.6s	−25° 16′ 16″	4.14
ω	20h 51m 49.2s	−26° 55′ 09″	4.11

Deep sky object	Description	Right ascension	Declination	Mag.
M30	Globular Cluster	21h 40m 21.9s	−23° 10.7′	7.3

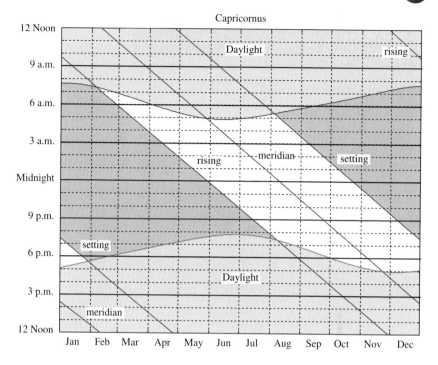

Capricornus

Carina

the Ship's Keel

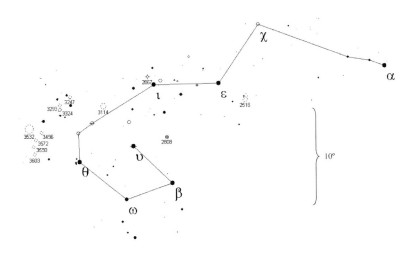

Star	Right ascension	Declination	Mag.
α	06h 23m 57.2s	−52° 41′ 44″	−0.72
β	09h 13m 12.1s	−69° 43′ 02″	1.68
ε	08h 22m 30.8s	−59° 30′ 34″	1.86
η	10h 45m 03.6s	−59° 41′ 03″	6.21
θ	10h 42m 57.4s	−64° 23′ 40″	2.76
ι	09h 17m 05.4s	−59° 16′ 31″	2.25
υ_1	09h 47m 06.1s	−65° 04′ 18″	2.96
υ_2	09h 47m 06.7s	−65° 04′ 21″	6.03
χ	07h 56m 46.7s	−52° 58′ 56″	3.47
ω	10h 13m 44.3s	−70° 02′ 16″	3.32

Deep sky object	Description	Right ascension	Declination	Mag.
NGC 2516	Open Cluster	07h 58m 07.2s	−60° 45.2′	3.8
NGC 2808	Globular Cluster	09h 12m 02.6s	−64° 51.8′	6.3
NGC 2867	Planetary Nebula	09h 21m 25.4s	−58° 18.7′	9.7
NGC 3114	Open Cluster	10h 02m 42.7s	−60° 06.5′	4.2
NGC 3247	Open Cluster	10h 25m 52.0s	−57° 55.6′	7.6
NGC 3293	Open Cluster	10h 35m 53.8s	−58° 14.2′	4.7
NGC 3324	Open Cluster	10h 37m 18.7s	−58° 39.6′	6.7
NGC 3496	Open Cluster	10h 59m 33.8s	−60° 20.2′	8.2
NGC 3532	Open Cluster	11h 05m 47.5s	−58° 46.2′	3.0
NGC 3572	Open Cluster	11h 10m 19.2s	−60° 14.9′	6.6
NGC 3590	Open Cluster	11h 12m 59.0s	−60° 47.3′	8.2
NGC 3603	Open Cluster	11h 15m 06.0s	−61° 15.0′	9.1

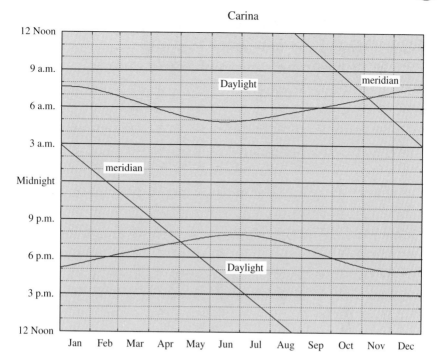

Carina

The constellation Carina is not visible from mid-northern latitudes

Cassiopeia
Queen of Ethiopia

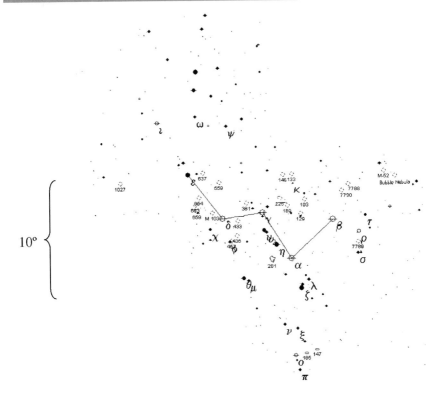

Star	Right ascension	Declination	Mag.
α	00h 40m 30.4s	+56° 32′ 15″	2.23
β	00h 09m 10.6s	+59° 08′ 59″	2.27
γ	00h 56m 42.4s	+60° 43′ 00″	2.47
δ	01h 25m 48.9s	+60° 14′ 07″	2.68
ε	01h 54m 23.6s	+63° 40′ 13″	3.38
ζ	00h 36m 58.2s	+53° 53′ 49″	3.66
η	00h 49m 06.0s	+57° 48′ 58″	3.44
θ	01h 11m 06.1s	+55° 09′ 00″	4.33
ι	02h 29m 03.9s	+67° 24′ 09″	4.52
κ	00h 32m 59.9s	+62° 55′ 55″	4.16
λ	00h 31m 46.3s	+54° 31′ 20″	4.73
μ	01h 08m 16.3s	+54° 55′ 14″	5.17
ν	00h 48m 50.0s	+50° 58′ 06″	4.89
ξ	00h 42m 03.8s	+50° 30′ 45″	4.80
ο	00h 44m 43.5s	+48° 17′ 04″	4.54
π	00h 43m 28.0s	+47° 01′ 29″	4.94
ρ	23h 54m 23.0s	+57° 29′ 58″	4.54
σ	23h 59m 00.4s	+55° 45′ 18″	4.88
τ	23h 47m 03.4s	+58° 39′ 07″	4.87
υ₁	00h 56m 39.7s	+59° 10′ 52″	4.63
υ₂	00h 55m 00.0s	+58° 58′ 22″	4.83
φ	01h 20m 04.8s	+58° 13′ 54″	4.98

Star	Right ascension	Declination	Mag.
χ	01h 33m 55.9s	+59° 13′ 56″	4.71
ψ	01h 25m 56.0s	+68° 07′ 48″	4.74
ω	01h 55m 59.9s	+68° 41′ 07″	4.99

Deep sky object	Description	Right ascension	Declination	Mag.
M52	Open Cluster	23h 24m 12.0s	+61° 35.0′	6.9
M103	Open Cluster	01h 33m 21.8s	+60° 39.5′	7.4
NGC 103	Open Cluster	00h 25m 17.4s	+61° 19.3′	9.8
NGC 129	Open Cluster	00h 29m 54.1s	+60° 12.6′	6.5
NGC 133	Open Cluster	00h 31m 16.9s	+63° 21.2′	8.5
NGC 146	Open Cluster	00h 33m 03.9s	+63° 18.6′	9.1
NGC 147	Galaxy	00h 33m 12.0s	+48° 30.0′	9.3
NGC 185	Galaxy	00h 39m 00.0s	+48° 20.0′	9.2
NGC 189	Open Cluster	00h 39m 35.7s	+61° 05.7′	8.8
NGC 225	Open Cluster	00h 43m 32.3s	+61° 47.4′	7.0
NGC 281	Open Cluster	00h 52m 59.3s	+56° 37.3′	7.4
NGC 381	Open Cluster	01h 08m 14.9s	+61° 35.0′	9.3
NGC 433	Open Cluster	01h 15m 09.2s	+60° 07.5′	9.0
NGC 436	Open Cluster	01h 15m 57.7s	+58° 49.0′	8.8
NGC 457	Open Cluster	01h 19m 32.6s	+58° 17.4′	6.4
NGC 559	Open Cluster	01h 29m 29.1s	+63° 18.5′	9.5
NGC 637	Open Cluster	01h 43m 03.1s	+64° 02.2′	8.2
NGC 654	Open Cluster	01h 43m 59.4s	+61° 53.0′	6.5
NGC 659	Open Cluster	01h 44m 23.0s	+60° 40.1′	7.9
NGC 663	Open Cluster	01h 46m 16.0s	+61° 13.1′	7.1
NGC 1027	Open Cluster	02h 42m 35.1s	+61° 35.7′	6.7
NGC 7635	Bubble Nebula	23h 20m 42.0s	+61° 12.0′	6.9
NGC 7788	Open Cluster	23h 56m 42.0s	+61° 24.0′	9.4
NGC 7789	Open Cluster	23h 57m 00.0s	+56° 44.0′	6.7
NGC 7790	Open Cluster	23h 58m 24.0s	+61° 13.0′	8.5

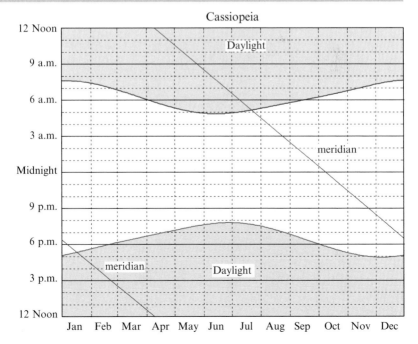

Cassiopeia

Centaurus

the Centaur

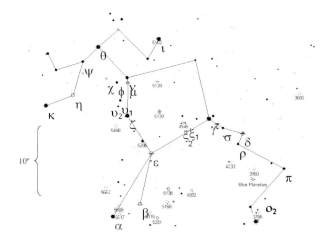

Star	Right ascension	Declination	Mag.
α_1	14h 39m 36.2s	−60° 50′ 07″	−0.01
α_2	14h 39m 36.2s	−60° 50′ 07″	1.33
β	14h 03m 49.4s	−60° 22′ 22″	0.61
γ	12h 41m 30.9s	−48° 57′ 34″	2.17
δ	12h 08m 21.5s	−50° 43′ 20″	2.60
ε	13h 39m 53.2s	−53° 27′ 59″	2.30
ζ	13h 55m 32.3s	−47° 17′ 18″	2.55
η	14h 35m 30.3s	−42° 09′ 28″	2.31
θ	14h 06m 40.8s	−36° 22′ 12″	2.06
ι	13h 20m 35.7s	−36° 42′ 44″	2.75
κ	14h 59m 09.6s	−42° 06′ 15″	3.13
λ	11h 35m 46.8s	−63° 01′ 11″	3.13
μ	13h 49m 36.9s	−42° 28′ 26″	3.04
ν	13h 49m 30.2s	−41° 41′ 16″	3.41
ξ_1	13h 03m 33.1s	−49° 31′ 38″	4.85
ξ_2	13h 06m 54.5s	−49° 54′ 22″	4.27
o_1	11h 31m 46.1s	−59° 26′ 32″	5.13
o_2	11h 31m 48.6s	−59° 30′ 56″	5.15
π	11h 21m 00.4s	−54° 29′ 27″	3.89
ρ	12h 11m 39.1s	−52° 22′ 06″	3.96
σ	12h 28m 02.3s	−50° 13′ 51″	3.91
τ	12h 37m 42.1s	−48° 32′ 28″	3.86
υ_1	13h 58m 40.7s	−44° 48′ 13″	3.87
υ_2	14h 01m 43.3s	−45° 36′ 12″	4.34
φ	13h 58m 16.2s	−42° 06′ 03″	3.83
χ	14h 06m 02.7s	−41° 10′ 46″	4.36
ψ	14h 20m 33.3s	−37° 53′ 07″	4.05

Deep sky object	Description	Right ascension	Declination	Mag.
NGC 3680	Open Cluster	11h 25m 42.0s	–43° 15.0′	7.6
NGC 3766	Open Cluster	11h 36m 06.0s	–61° 37.0′	5.3
NGC 3918	Blue Planetary	11h 50m 17.8s	–57° 10.9′	8.0
NGC 3960	Open Cluster	11h 50m 33.2s	–55° 40.6′	8.3
NGC 4230	Open Cluster	12h 17m 18.0s	–55° 08.0′	9.4
NGC 4852	Open Cluster	13h 00m 06.0s	–59° 36.0′	8.9
NGC 4945	Galaxy	13h 05m 24.0s	–49° 28.0′	9.6
NGC 5102	Galaxy	13h 21m 57.5s	–36° 37.8′	9.7
NGC 5128	Peculiar Galaxy	13h 25m 29.0s	–43° 01.0′	7.0
NGC 5138	Open Cluster	13h 27m 15.2s	–59° 02.5′	7.6
NGC 5139	ω Centauri Globular Cluster	13h 26m 45.9s	–47° 28.6′	3.5
NGC 5168	Open Cluster	13h 31m 07.3s	–60° 56.3′	9.1
NGC 5281	Open Cluster	13h 46m 36.0s	–62° 54.0′	5.9
NGC 5286	Globular Cluster	13h 46m 24.0s	–51° 22.0′	7.6
NGC 5316	Open Cluster	13h 53m 54.0s	–61° 52.0′	6.0
NGC 5460	Open Cluster	14h 07m 36.0s	–48° 19.0′	5.6
NGC 5606	Open Cluster	14h 27m 48.0s	–59° 38.0′	7.7
NGC 5617	Open Cluster	14h 29m 48.0s	–60° 43.0′	6.3
NGC 5662	Open Cluster	14h 35m 12.0s	–56° 33.0′	5.5

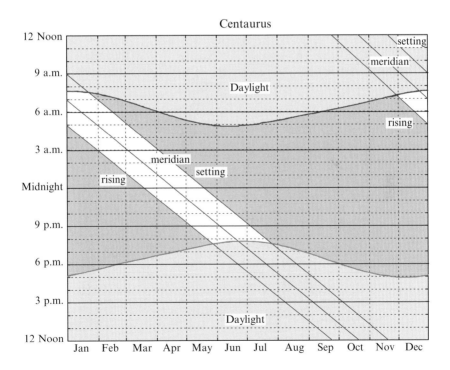

Centaurus

Cepheus

King of Ethiopia

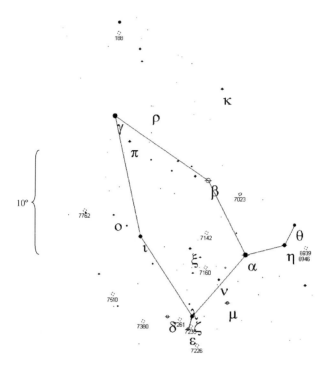

Star	Right ascension	Declination	Mag.
α	21h 18m 34.7s	+62° 35′ 08″	2.44
β	21h 28m 39.5s	+70° 33′ 39″	3.23
γ	23h 39m 20.8s	+77° 37′ 57″	3.21
δ	22h 29m 10.2s	+58° 24′ 55″	3.75
ε	22h 15m 01.9s	+57° 02′ 37″	4.19
ζ	22h 10m 51.2s	+58° 12′ 05″	3.35
η	20h 45m 17.3s	+61° 50′ 20″	3.43
θ	20h 29m 34.8s	+62° 59′ 39″	4.22
ι	22h 49m 40.7s	+66° 12′ 02″	3.52
κ	20h 08m 53.2s	+77° 42′ 41″	4.39
λ	22h 11m 30.6s	+59° 24′ 53″	5.04
μ	21h 43m 30.3s	+58° 46′ 48″	4.08
ν	21h 45m 26.8s	+61° 07′ 15″	4.29
ξ	22h 03m 47.3s	+64° 37′ 41″	4.29
o	23h 18m 37.4s	+68° 06′ 42″	4.75
π	23h 07m 53.8s	+75° 23′ 16″	4.41
ρ₁	22h 26m 42.4s	+78° 47′ 09″	5.83
ρ₂	22h 29m 52.8s	+78° 49′ 28″	5.52

Deep sky object	Description	Right ascension	Declination	Mag.
NGC 188	Open Cluster	00h 47m 29.7s	+85° 14.5'	8.1
NGC 6939	Open Cluster	20h 31m 24.0s	+60° 38.0'	7.8
NGC 6946	Spiral Galaxy	20h 34m 48.0s	+60° 09.0'	8.8
NGC 7023	Open Cluster	21h 01m 35.5s	+68° 10.2'	7.1
NGC 7142	Open Cluster	21h 45m 09.4s	+65° 46.5'	9.3
NGC 7160	Open Cluster	21h 53m 40.2s	+62° 36.2'	6.1
NGC 7226	Open Cluster	22h 10m 30.0s	+55° 25.0'	9.6
NGC 7235	Open Cluster	22h 12m 36.0s	+57° 17.0'	7.7
NGC 7261	Open Cluster	22h 20m 24.0s	+58° 05.0'	8.4
NGC 7380	Open Cluster	22h 47m 00.0s	+58° 06.0'	7.2
NGC 7510	Open Cluster	23h 11m 30.0s	+60° 34.0'	7.9
NGC 7762	Open Cluster	23h 49m 48.0s	+68° 02.0'	10.0

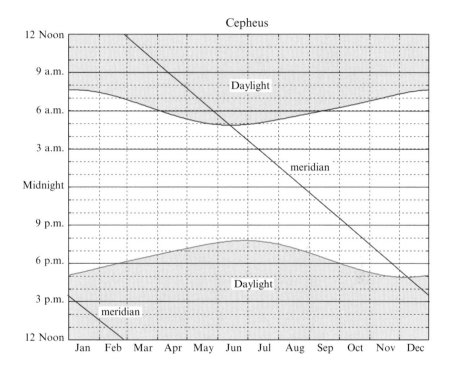

Cepheus

Cetus

the Whale

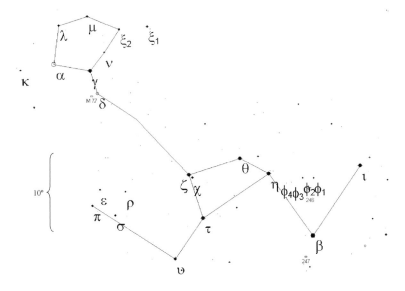

Star	Right ascension	Declination	Mag.
α	03h 02m 16.7s	+04° 05′ 23″	2.53
β	00h 43m 35.3s	−17° 59′ 12″	2.04
γ	02h 43m 18.0s	+03° 14′ 09″	3.47
δ	02h 39m 28.9s	+00° 19′ 43″	4.07
ε	02h 39m 33.7s	−11° 52′ 20″	4.84
ζ	01h 51m 27.5s	−10° 20′ 06″	3.73
η	01h 08m 35.3s	−10° 10′ 56″	3.45
θ	01h 24m 01.3s	−08° 11′ 01″	3.60
ι	00h 19m 25.6s	−08° 49′ 26″	3.56
κ₁	03h 19m 21.6s	+03° 22′ 13″	4.83
κ₂	03h 21m 06.7s	+03° 40′ 32″	5.69
λ	02h 59m 42.8s	+08° 54′ 27″	4.70
μ	02h 44m 56.5s	+10° 06′ 51″	4.27
ν	02h 35m 52.4s	+05° 35′ 36″	4.86
ξ₁	02h 12m 59.9s	+08° 50′ 48″	4.37
ξ₂	02h 28m 09.5s	+08° 27′ 36″	4.28
o	02h 19m 20.7s	−02° 58′ 39″	3.04
π	02h 44m 07.3s	−13° 51′ 32″	4.25
ρ	02h 25m 56.9s	−12° 17′ 26″	4.89
σ	02h 32m 05.1s	−15° 14′ 41″	4.75
τ	01h 44m 04.0s	−15° 56′ 15″	3.50
υ	02h 00m 00.2s	−21° 04′ 40″	4.00
φ₁	00h 44m 11.3s	−10° 36′ 34″	4.76
φ₂	00h 50m 07.5s	−10° 38′ 40″	5.19
φ₃	00h 56m 01.4s	−11° 16′ 00″	5.31
φ₄	00h 58m 43.8s	−11° 22′ 49″	5.61
χ	01h 49m 35.0s	−10° 41′ 11″	4.67

Deep sky object	Description	Right ascension	Declination	Mag.
M77	Galaxy	02h 42m 40.7s	–00° 00.8′	8.8
NGC 246	Planetary Nebula	00h 47m 00.0s	–11° 53.0′	8.0
NGC 247	Galaxy	00h 47m 08.4s	–20° 45.6′	9.1

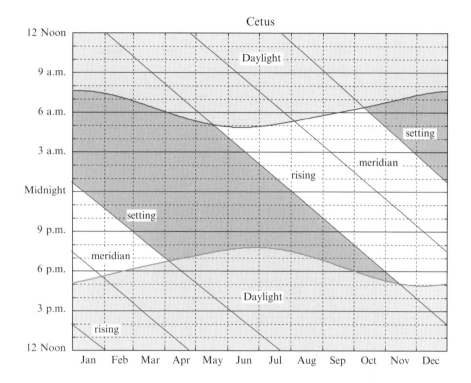
Cetus

Chamaeleon
the Chameleon

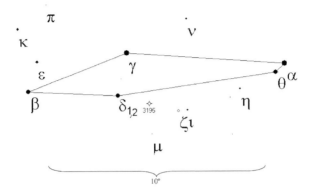

Star	Right ascension	Declination	Mag.
α	08h 18m 31.7s	−76° 55′ 11″	4.07
β	12h 18m 20.7s	−79° 18′ 43″	4.26
γ	10h 35m 28.1s	−78° 36′ 27″	4.11
δ₁	10h 45m 15.8s	−80° 28′ 10″	5.47
δ₂	10h 45m 46.6s	−80° 32′ 24″	4.45
ε	11h 59m 37.3s	−78° 13′ 18″	4.91
ζ	09h 33m 53.4s	−80° 56′ 29″	5.11
η	08h 41m 19.9s	−78° 57′ 48″	5.47
θ	08h 20m 38.7s	−77° 29′ 04″	4.35
ι	09h 24m 09.2s	−80° 47′ 13″	5.36
κ	12h 04m 46.5s	−76° 31′ 08″	5.04
μ₁	10h 00m 43.7s	−82° 12′ 52″	5.52
μ₂	10h 04m 07.7s	−81° 33′ 56″	6.60
ν	09h 46m 21.3s	−76° 46′ 33″	5.45
π	11h 37m 15.6s	−75° 53′ 47″	5.65

Deep sky object	Description	Right ascension	Declination	Mag.
NGC 3195	Planetary Nebula	10h 09m 21.1s	−80° 51.5′	10.0

Chamaeleon

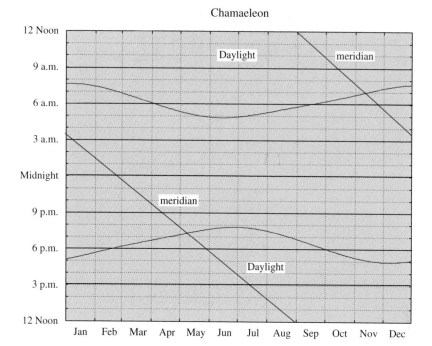

The constellation Chamaeleon is not visible from mid-northern latitudes

Circinus

the Compass

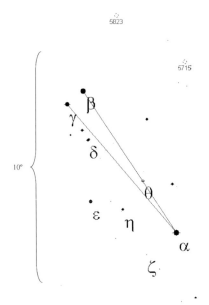

Star	Right ascension	Declination	Mag.
α	14h 42m 30.3s	−64° 58′ 31″	3.19
β	15h 17m 30.8s	−58° 48′ 04″	4.07
γ	15h 23m 22.6s	−59° 19′ 15″	4.51
δ	15h 16m 56.7s	−60° 57′ 27″	5.09
ε	15h 17m 38.8s	−63° 36′ 38″	4.86
ζ	14h 54m 42.4s	−65° 59′ 29″	6.09
η	15h 04m 48.1s	−64° 01′ 54″	5.17
θ	14h 56m 44.0s	−62° 46′ 51″	5.11

Deep sky object	Description	Right ascension	Declination	Mag.
NGC 5715	Open Cluster	14h 43m 24.0s	−57° 33.0′	9.8
NGC 5823	Open Cluster	15h 05m 42.0s	−55° 36.0′	7.9

Circinus

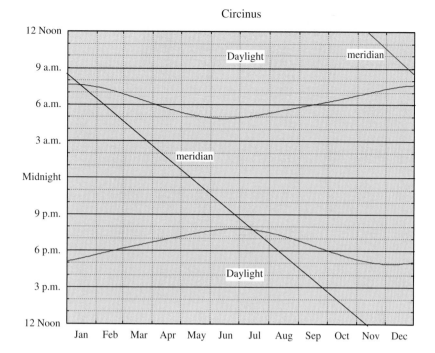

The constellation Circinus is not visible from mid-northern latitudes

Columba

the Dove

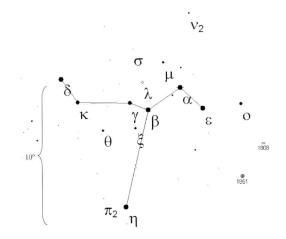

Star	Right ascension	Declination	Mag.
α	05h 39m 38.9s	−34° 04′ 27″	2.64
β	05h 50m 57.5s	−35° 46′ 06″	3.12
γ	05h 57m 32.2s	−35° 17′ 00″	4.36
δ	06h 22m 06.7s	−33° 26′ 11″	3.85
ε	05h 31m 12.7s	−35° 28′ 14″	3.87
η	05h 59m 08.7s	−42° 48′ 55″	3.96
θ	06h 07m 31.6s	−37° 15′ 10″	5.02
κ	06h 16m 33.0s	−35° 08′ 26″	4.37
λ	05h 53m 06.8s	−33° 48′ 05″	4.87
μ	05h 45m 59.9s	−32° 18′ 23″	5.17
ν_1	05h 37m 16.5s	−27° 52′ 17″	6.16
ν_2	05h 37m 44.6s	−28° 41′ 22″	5.31
ξ	05h 55m 29.8s	−37° 07′ 15″	4.97
o	05h 17m 29.0s	−34° 53′ 43″	4.83
π_1	06h 06m 41.0s	−42° 17′ 55″	6.12
π_2	06h 07m 52.9s	−42° 09′ 14″	5.50
σ	05h 56m 20.9s	−31° 22′ 57″	5.50

Deep sky object	Description	Right ascension	Declination	Mag.
NGC 1808	Galaxy	05h 07m 42.3s	−37° 30.8′	9.9
NGC 1851	Globular Cluster	05h 14m 06.3s	−40° 02.8′	7.2

Columba

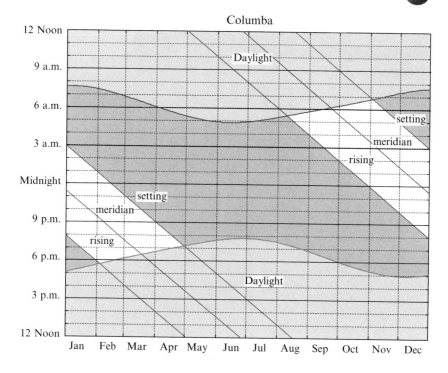

Coma Berenices

the Hair of Queen Berenice of Egypt

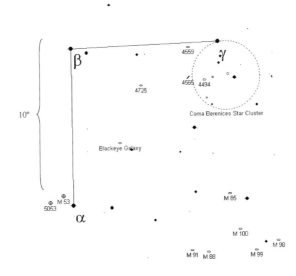

Star	Right ascension	Declination	Mag.
α_1	13h 09m 59.2s	+17° 31′ 46″	5.22
α_2	13h 09m 59.2s	+17° 31′ 46″	5.22
β	13h 11m 52.3s	+27° 52′ 41″	4.26
γ	12h 26m 56.2s	+28° 16′ 06″	4.36

Deep sky object	Description	Right ascension	Declination	Mag.
M53	Globular Cluster	13h 12m 55.2s	+18° 10.1′	7.5
M64	The Black Eye Galaxy	12h 56m 42.0s	+21° 41.0′	8.5
M85	Galaxy	12h 25m 24.2s	+18° 11.5′	9.1
M88	Galaxy	12h 31m 59.1s	+14° 25.2′	9.5
M91	Galaxy	12h 35m 27.2s	+14° 29.8′	10.2
M98	Galaxy	12h 13m 48.2s	+14° 54.0′	10.1
M99	Galaxy	12h 18m 49.6s	+14° 25.0′	9.9
M100	Galaxy	12h 22m 54.9s	+15° 49.3′	9.4
NGC 4494	Galaxy	12h 31m 24.3s	+25° 46.5′	9.9
NGC 4559	Spiral Galaxy	12h 35m 57.7s	+27° 57.6′	9.9
NGC 4565	Spiral Galaxy	12h 36m 20.7s	+25° 59.3′	9.5
NGC 4725	Galaxy	12h 50m 24.0s	+25° 30.0′	9.2
NGC 5053	Globular Cluster	13h 16m 27.0s	+17° 41.9′	9.9
Melotte 111	Coma Berenices Star Cluster	12h 25m 00.0s	+26° 00.0′	1.8

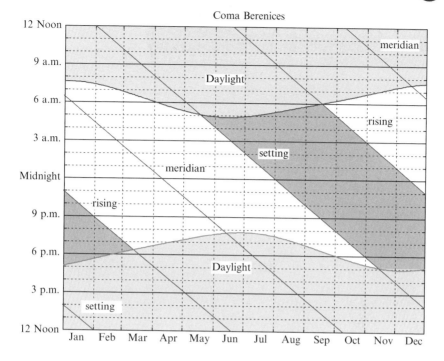

Coma Berenices

Corona Australis
the Southern Crown

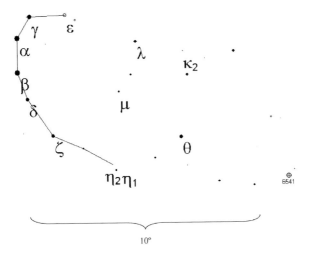

10°

Star	Right ascension	Declination	Mag.
α	19h 09m 28.2s	−37° 54′ 16″	4.11
β	19h 10m 01.6s	−39° 20′ 27″	4.11
γ_1	19h 56m 25.0s	−37° 03′ 48″	5.01
γ_2	19h 56m 25.0s	−37° 03′ 48″	5.01
δ	19h 58m 20.8s	−40° 29′ 48″	4.59
ϵ	18h 58m 43.3s	−37° 06′ 26″	4.87
ζ	19h 53m 56.7s	−42° 05′ 43″	4.75
η_1	18h 48m 50.4s	−43° 40′ 48″	5.49
η_2	18h 49m 34.9s	−43° 26′ 02″	5.61
θ	18h 33m 30.1s	−42° 18′ 45″	4.64
κ_1	18h 33m 23.2s	−38° 43′ 13″	6.32
κ_2	18h 33m 23.0s	−38° 43′ 34″	5.65
λ	18h 43m 46.8s	−38° 19′ 25″	5.13
μ	18h 47m 44.4s	−40° 24′ 22″	5.24

Deep sky object	Description	Right ascension	Declination	Mag.
NGC 6541	Globular Cluster	18h 08m 00.0s	−43° 42.0′	6.1

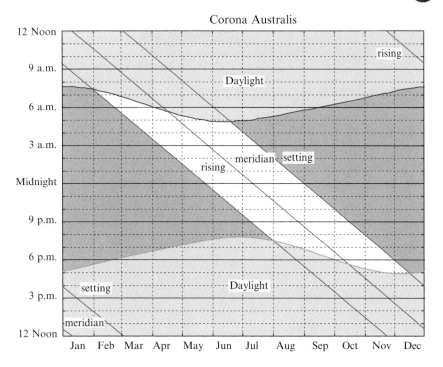
Corona Australis

Corona Borealis
the Northern Crown

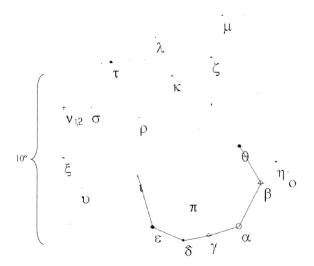

Star	Right ascension	Declination	Mag.
α	15h 34m 41.2s	+26° 42′ 53″	2.23
β	15h 27m 49.7s	+29° 06′ 20″	3.68
γ	15h 42m 44.5s	+26° 17′ 44″	3.84
δ	15h 49m 35.6s	+26° 04′ 06″	4.63
ε	15h 57m 35.2s	+26° 52′ 40″	4.15
ζ_1	15h 39m 22.1s	+36° 38′ 12″	6.00
ζ_2	15h 39m 22.6s	+36° 38′ 09″	5.07
η_1	15h 23m 12.2s	+30° 17′ 16″	5.58
η_2	15h 23m 12.2s	+30° 17′ 16″	6.08
θ	15h 32m 55.7s	+31° 21′ 32″	4.14
ι	16h 01m 26.6s	+29° 51′ 04″	4.99
κ	15h 51m 13.8s	+35° 39′ 26″	4.82
λ	15h 55m 47.6s	+37° 56′ 49″	5.45
μ	15h 35m 14.8s	+39° 00′ 36″	5.11
ν_1	16h 22m 21.3s	+33° 47′ 56″	5.20
ν_2	16h 22m 29.1s	+33° 42′ 13″	5.39
ξ	16h 22m 05.7s	+30° 53′ 32″	4.85
o	15h 20m 08.4s	+29° 36′ 58″	5.51
π	15h 43m 59.2s	+32° 30′ 57″	5.56
ρ	16h 01m 02.6s	+33° 18′ 13″	5.41
σ_1	16h 14m 40.7s	+33° 51′ 30″	5.64
σ_2	16h 14m 40.7s	+33° 51′ 30″	6.66
τ	16h 08m 58.2s	+36° 29′ 27″	4.76
υ	16h 16m 44.7s	+29° 09′ 01″	5.78

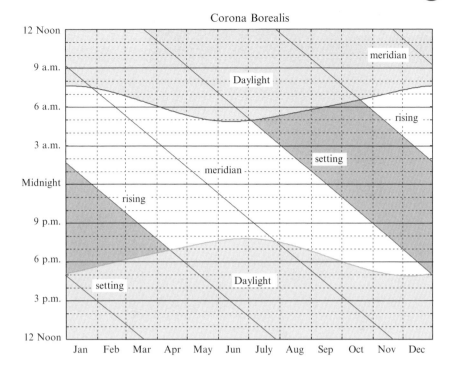

Corona Borealis

Corvus
the Crow

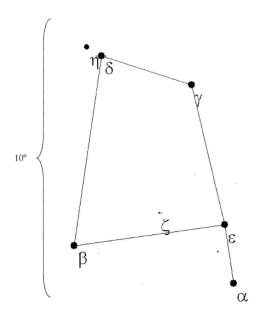

Star	Right ascension	Declination	Mag.
α	12h 08m 24.7s	−24° 43′ 44″	4.02
β	12h 34m 23.2s	−23° 23′ 48″	2.65
γ	12h 15m 48.3s	−17° 32′ 31″	2.59
δ	12h 29m 51.8s	−16° 30′ 56″	2.95
ε	12h 10m 07.4s	−22° 37′ 11″	3.00
ζ	12h 20m 33.6s	−22° 12′ 57″	5.21
η	12h 32m 04.1s	−16° 11′ 46″	4.31

Corvus

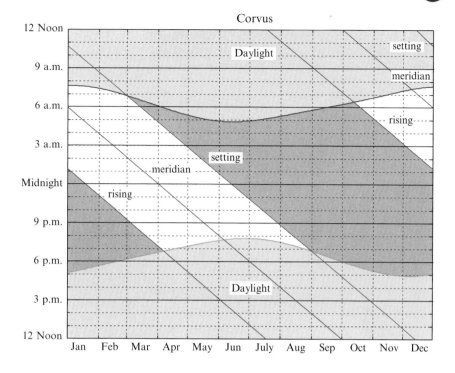

Crater

the Cup

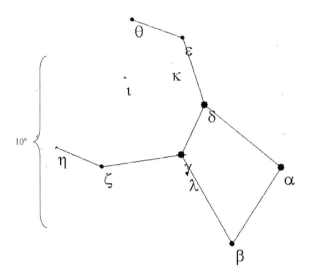

Star	Right ascension	Declination	Mag.
α	10h 59m 46.4s	−18° 17′ 56″	4.08
β	11h 11m 39.4s	−22° 49′ 33″	4.48
γ	11h 24m 52.8s	−17° 41′ 03″	4.08
δ	11h 19m 20.4s	−14° 46′ 43″	3.56
ε	11h 24m 36.5s	−10° 51′ 34″	4.83
ζ	11h 44m 45.7s	−18° 21′ 03″	4.73
η	11h 56m 00.9s	−17° 09′ 03″	5.18
θ	11h 36m 40.8s	−09° 48′ 08″	4.70
ι	11h 38m 40.0s	−13° 12′ 07″	5.48
κ	11h 27m 09.4s	−12° 21′ 25″	5.94
λ	11h 23m 21.8s	−18° 46′ 48″	5.09

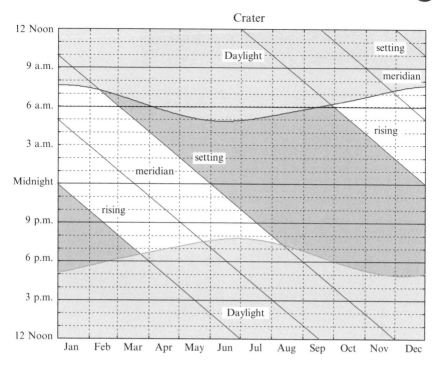

Crater

Crux

the Southern Cross

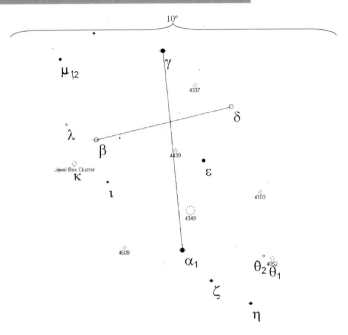

Star	Right ascension	Declination	Mag.
α_1	12h 26m 35.9s	−63° 05′ 56″	1.58
α_2	12h 26m 36.5s	−63° 05′ 58″	2.09
β	12h 47m 43.3s	−59° 41′ 19″	1.25
γ_1	12h 31m 09.9s	−57° 06′ 47″	1.63
γ_2	12h 31m 16.7s	−57° 04′ 51″	6.42
δ	12h 15m 08.6s	−58° 44′ 56″	2.80
ε	12h 21m 21.5s	−60° 24′ 04″	3.59
ζ	12h 18m 26.1s	−64° 00′ 11″	4.04
η	12h 06m 52.8s	−64° 36′ 49″	4.15
θ_1	12h 03m 01.5s	−63° 18′ 46″	4.33
θ_2	12h 04m 19.2s	−63° 09′ 56″	4.72
ι	12h 45m 37.8s	−60° 58′ 52″	4.69
κ	12h 53m 49.0s	−60° 22′ 36″	5.90
λ	12h 54m 39.1s	−59° 08′ 47″	4.62
μ_1	12h 54m 35.6s	−57° 10′ 40″	4.03
μ_2	12h 54m 36.8s	−57° 10′ 05″	5.17

Deep sky object	Description	Right ascension	Declination	Mag.
NGC 4052	Open Cluster	12h 02m 05.2s	−63° 13.4′	8.8
NGC 4103	Open Cluster	12h 06m 39.5s	−61° 15.0′	7.4
NGC 4337	Open Cluster	12h 24m 03.3s	−58° 07.4′	8.9
NGC 4349	Open Cluster	12h 24m 06.0s	−61° 52.2′	7.4
NGC 4439	Open Cluster	12h 28m 26.3s	−60° 06.2′	8.4
NGC 4609	Open Cluster	12h 42m 18.0s	−62° 58.0′	6.9
NGC 4755	Jewel Box Cluster	12h 53m 36.0s	−60° 20.0′	4.2

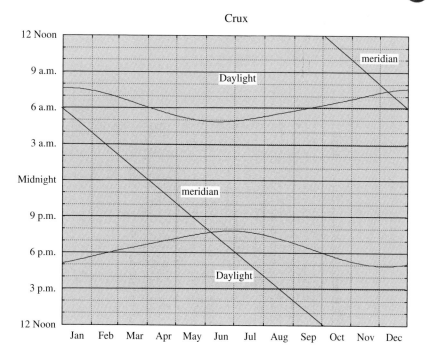

Crux

The constellation Crux is not visible from mid-northern latitudes

Cygnus

the Swan

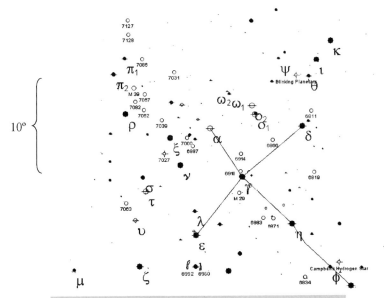

Star	Right ascension	Declination	Mag.
α	20h 41m 25.8s	+45° 16′ 49″	1.25
β₁	19h 30m 43.2s	+27° 57′ 35″	3.08
β₂	19h 30m 45.2s	+27° 57′ 55″	5.11
γ	20h 22m 13.6s	+40° 15′ 24″	2.20
δ	19h 44m 58.4s	+45° 07′ 51″	2.87
ε	20h 46m 12.6s	+33° 58′ 13″	2.46
ζ	21h 12m 56.1s	+30° 13′ 37″	3.20
η	19h 56m 18.3s	+35° 05′ 00″	3.89
θ	19h 36m 26.4s	+50° 13′ 16″	4.48
ι	19h 29m 42.2s	+51° 43′ 47″	3.79
κ	19h 17m 06.0s	+53° 22′ 07″	3.77
λ	20h 47m 24.4s	+36° 29′ 27″	4.53
μ₁	21h 44m 08.5s	+28° 44′ 34″	4.73
μ₂	21h 44m 08.2s	+28° 44′ 35″	6.08
ν	20h 57m 10.3s	+41° 10′ 02″	3.94
ξ	21h 04m 55.8s	+43° 55′ 40″	3.72
π₁	21h 46m 47.5s	+49° 18′ 35″	4.23
π₂	21h 42m 05.6s	+51° 11′ 23″	4.67
ρ	21h 33m 58.8s	+45° 35′ 31″	4.02
σ	21h 17m 24.9s	+39° 23′ 41″	4.23
τ	21h 14m 47.4s	+38° 02′ 44″	3.72
υ	21h 17m 55.0s	+34° 53′ 49″	4.43
φ	19h 39m 22.5s	+30° 09′ 12″	4.69
χ	19h 50m 33.8s	+32° 54′ 51″	4.23
ψ	19h 55m 37.7s	+52° 26′ 20″	4.92
ω₁	20h 30m 03.4s	+48° 57′ 06″	4.95
ω₂	20h 31m 18.7s	+49° 13′ 14″	5.44

Deep sky object	Description	Right ascension	Declination	Mag.
M29	Open Cluster	20h 23m 54.0s	+38° 32.0′	6.6
M39	Open Cluster	21h 31m 48.3s	+48° 26.9′	4.6
NGC 6811	Open Cluster	19h 38m 12.0s	+46° 34.0′	6.8
NGC 6819	Open Cluster	19h 41m 18.0s	+40° 11.0′	7.3
NGC 6826	Blinking Planetary Nebula	19h 44m 48.2s	+50° 31.5′	8.8
NGC 6834	Open Cluster	19h 52m 12.0s	+29° 25.0′	7.8
NGC 6866	Open Cluster	20h 03m 42.0s	+44° 00.0′	7.6
NGC 6871	Open Cluster	20h 05m 54.0s	+35° 47.0′	5.2
NGC 6883	Open Cluster	20h 11m 18.0s	+35° 51.0′	8.0
NGC 6910	Open Cluster	20h 23m 06.0s	+40° 47.0′	7.4
NGC 6914	Diffuse Nebula	20h 24m 42.0s	+42° 29.0′	10.0
NGC 6960	Veil Nebula	20h 45m 42.0s	+30° 43.0′	7.0
NGC 6992	Veil Nebula	20h 56m 24.0s	+31° 43.0′	7.0
NGC 6997	Open Cluster	20h 56m 24.0s	+45° 28.0′	10.0
NGC 7000	North American Nebula	21h 01m 48.0s	+44° 12.0′	4.5
NGC 7027	Planetary Nebula	21h 07m 01.7s	+42° 14.2′	8.5
NGC 7031	Open Cluster	21h 07m 12.5s	+50° 52.5′	9.1
NGC 7039	Open Cluster	21h 10m 47.7s	+45° 37.3′	7.6
NGC 7062	Open Cluster	21h 23m 27.4s	+46° 22.7′	8.3
NGC 7063	Open Cluster	21h 24m 21.7s	+36° 29.2′	7.0
NGC 7067	Open Cluster	21h 24m 23.1s	+48° 00.6′	9.7
NGC 7082	Open Cluster	21h 29m 17.7s	+47° 07.6′	7.2
NGC 7086	Open Cluster	21h 30m 27.5s	+51° 36.1′	8.4
NGC 7127	Open Cluster	21h 43m 41.9s	+54° 37.8′	10.0
NGC 7128	Open Cluster	21h 43m 57.7s	+53° 42.9′	9.7

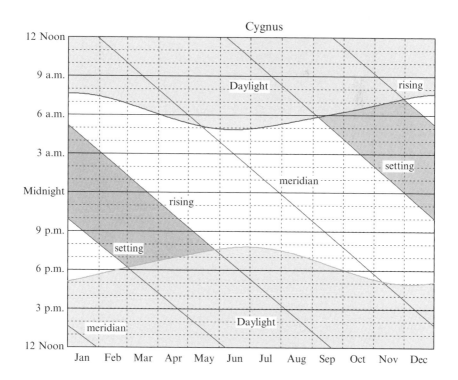

Cygnus

Delphinus
the Dolphin

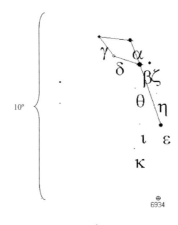

Star	Right ascension	Declination	Mag.
α	20h 39m 38.2s	+15° 54′ 43″	3.77
β	20h 37m 32.9s	+14° 35′ 43″	3.63
γ₁	20h 46m 38.6s	+16° 07′ 28″	5.14
γ₂	20h 46m 39.4s	+16° 07′ 27″	4.27
δ	20h 43m 27.5s	+15° 04′ 28″	4.43
ε	20h 33m 12.7s	+11° 18′ 12″	4.03
ζ	20h 35m 18.4s	+14° 40′ 27″	4.68
η	20h 33m 56.9s	+13° 01′ 38″	5.38
θ	20h 38m 43.8s	+13° 18′ 54″	5.72
ι	20h 37m 49.0s	+11° 22′ 40″	5.43
κ	20h 39m 07.7s	+10° 05′ 10″	5.05

Deep sky object	Description	Right ascension	Declination	Mag.
NGC 6934	Globular Cluster	20h 34m 12.0s	+07° 24.0′	8.7

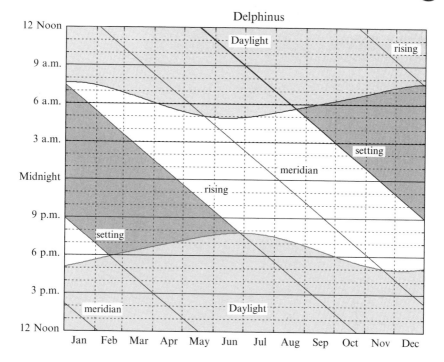

Delphinus

Dorado

the Swordfish

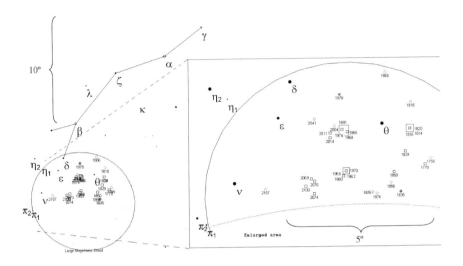

Star	Right ascension	Declination	Mag.
α	04h 33m 59.8s	−55° 02′ 42″	3.27
β	05h 33m 37.5s	−62° 29′ 24″	3.76
γ	04h 16m 01.6s	−51° 29′ 12″	4.25
δ	05h 44m 46.5s	−65° 44′ 08″	4.35
ϵ	05h 49m 53.7s	−66° 54′ 05″	5.11
ζ	05h 05m 30.6s	−57° 28′ 22″	4.72
η_1	06h 06m 09.5s	−66° 02′ 23″	5.71
η_2	06h 11m 15.0s	−65° 35′ 22″	5.01
θ	05h 13m 45.4s	−67° 11′ 08″	4.83
κ	04h 44m 21.2s	−59° 43′ 58″	5.27
λ	05h 26m 19.2s	−58° 54′ 46″	5.14
ν	06h 08m 44.3s	−68° 50′ 36″	5.06
π_1	06h 22m 38.3s	−69° 59′ 03″	5.56
π_2	06h 25m 28.7s	−69° 41′ 25″	5.38

Deep sky object	Description	Right ascension	Declination	Mag.
NGC 1755	Open Cluster	04h 55m 14.9s	−68° 12.3′	9.9
NGC 1770	Open Cluster	04h 57m 15.8s	−68° 25.1′	9.0
NGC 1814	Open Cluster	05h 03m 46.5s	−67° 18.1′	9.0
NGC 1816	Open Cluster	05h 03m 50.8s	−67° 15.7′	9.0
NGC 1818	Open Cluster	05h 04m 14.8s	−66° 26.1′	9.9
NGC 1820	Open Cluster	05h 04m 01.7s	−67° 16.0′	9.0
NGC 1829	Open Cluster	05h 04m 57.3s	−68° 03.3′	8.5
NGC 1835	Globular Cluster	05h 05m 05.7s	−69° 24.3′	9.8
NGC 1850	Globular Cluster	05h 08m 44.8s	−68° 45.7′	9.4
NGC 1856	Globular Cluster	05h 09m 29.5s	−69° 07.6′	10.0
NGC 1866	Open Cluster	05h 13m 39.1s	−65° 27.9′	9.9
NGC 1874	Open Cluster	05h 13m 11.7s	−69° 22.6′	9.0
NGC 1876	Open Cluster	05h 13m 18.5s	−69° 21.9′	9.0

Deep sky object	Description	Right ascension	Declination	Mag.
NGC 1955	Open Cluster	05h 26m 10.0s	−67° 29.9′	9.0
NGC 1962	Open Cluster	05h 26m 17.8s	−68° 50.3′	8.5
NGC 1965	Open Cluster	05h 26m 29.1s	−68° 48.4′	8.5
NGC 1968	Open Cluster	05h 27m 22.2s	−67° 27.8′	9.0
NGC 1970	Open Cluster	05h 26m 52.7s	−68° 50.2′	8.5
NGC 1974	Open Cluster	05h 28m 00.4s	−67° 25.4′	9.0
NGC 1978	Globular Cluster	05h 28m 45.3s	−66° 14.2′	9.9
NGC 1983	Open Cluster	05h 27m 44.3s	−68° 59.2′	8.5
NGC 1991	Open Cluster	05h 28m 00.4s	−67° 25.4′	9.0
NGC 2004	Open Cluster	05h 30m 40.3s	−67° 17.2′	9.9
NGC 2011	Open Cluster	05h 32m 20.3s	−67° 31.4′	9.5
NGC 2014	Open Cluster	05h 32m 19.9s	−67° 41.4′	8.5
NGC 2041	Open Cluster	05h 36m 28.1s	−66° 59.4′	10.0
NGC 2069	Diffuse Nebula	05h 38m 46.5s	−68° 58.5′	5.0
NGC 2070	Tarantula Nebula	05h 38m 38.4s	−69° 05.6′	8.2
NGC 2074	Diffuse Nebula	05h 39m 03.7s	−69° 29.9′	8.5
NGC 2100	Open Cluster	05h 42m 09.1s	−69° 12.7′	9.6
NGC 2157	Open Cluster	05h 57m 34.9s	−69° 11.8′	10.0
Large Magellanic Cloud		05h 23m 36.0s	−69° 45.0′	−0.1

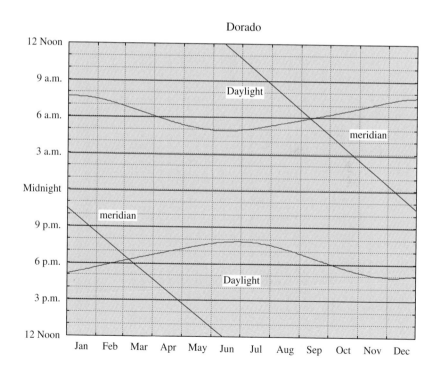

The constellation Dorado is not visible from mid-northern latitudes

Draco

the Dragon

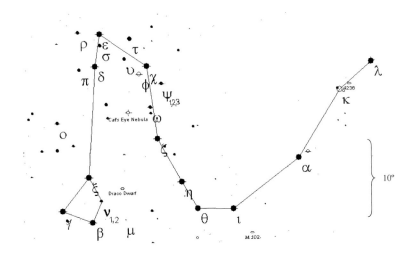

Star	Right ascension	Declination	Mag.
α	14h 04m 23.2s	+64° 22′ 33″	3.65
β	17h 30m 25.8s	+52° 18′ 05″	2.79
γ	17h 56m 36.3s	+51° 29′ 20″	2.23
δ	19h 12m 33.1s	+67° 39′ 42″	3.07
ε	19h 48m 10.3s	+70° 16′ 04″	3.83
ζ	17h 08m 47.1s	+65° 42′ 53″	3.17
η	16h 23m 59.3s	+61° 30′ 51″	2.74
θ	16h 01m 53.2s	+58° 33′ 55″	4.01
ι	15h 24m 55.6s	+58° 57′ 57″	3.29
κ	12h 33m 28.9s	+69° 47′ 17″	3.87
λ	11h 31m 24.2s	+69° 19′ 52″	3.84
μ_1	17h 05m 19.6s	+54° 28′ 13″	5.83
μ_2	17h 05m 19.6s	+54° 28′ 13″	5.80
ν_1	17h 32m 10.4s	+55° 11′ 03″	4.88
ν_2	17h 32m 15.9s	+55° 10′ 22″	4.87
ξ	17h 53m 31.6s	+56° 52′ 21″	3.75
o	18h 51m 11.9s	+59° 23′ 18″	4.66
π	19h 20m 40.0s	+65° 42′ 52″	4.59
ρ	20h 02m 48.9s	+67° 52′ 25″	4.51
σ	19h 32m 21.5s	+69° 39′ 40″	4.68
τ	19h 15m 32.8s	+73° 21′ 20″	4.45
υ	18h 54m 23.7s	+71° 17′ 50″	4.82
φ	18h 20m 45.3s	+71° 20′ 16″	4.22
χ	18h 21m 03.2s	+72° 43′ 58″	3.57
ψ_1	17h 41m 56.2s	+72° 08′ 56″	4.58
ψ_2	17h 41m 57.8s	+72° 09′ 25″	5.79
ψ_3	17h 55m 10.9s	+72° 00′ 19″	5.48
ω	17h 36m 56.9s	+68° 45′ 29″	4.80

Deep sky object	Description	Right ascension	Declination	Mag.
M102	Galaxy	15h 06m 30.0s	+55° 46.0′	10.0
NGC 4236	Galaxy	12h 16m 42.1s	+69° 27.7′	9.6
NGC 6543	The Cat's Eye Nebula	17h 58m 33.4s	+66° 38.0′	8.1

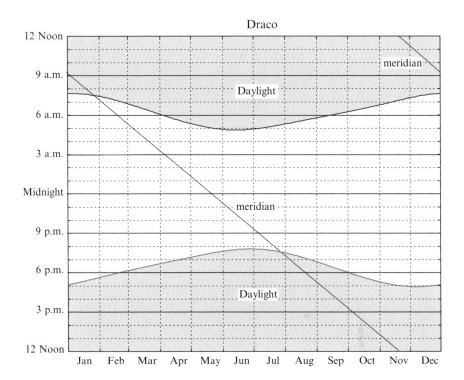

Equuleus

the Foal

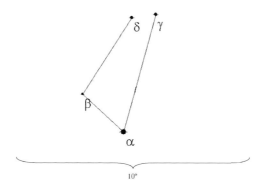

Star	Right ascension	Declination	Mag.
α	21h 15m 49.3s	+05° 14′ 52″	3.92
β	21h 22m 53.5s	+06° 48′ 40″	5.16
γ	21h 10m 20.5s	+10° 07′ 53″	4.69
δ	21h 14m 28.8s	+10° 00′ 25″	4.49
ε	20h 59m 04.3s	+04° 17′ 37″	5.23

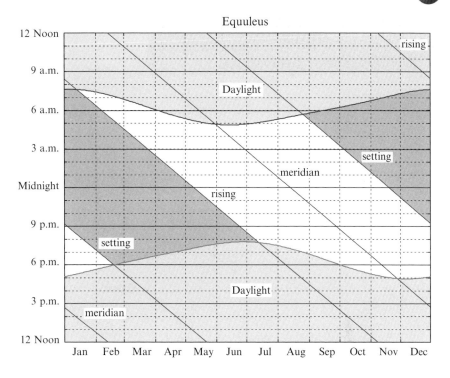

Equuleus

Eridanus

the River

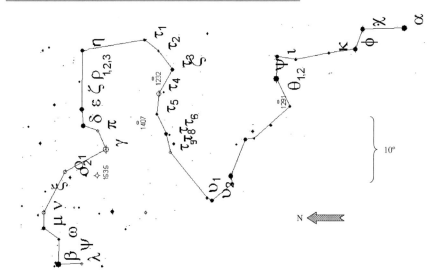

Star	Right ascension	Declination	Mag.
α	1h 37m 42.9s	−57° 14′ 12″	0.46
β	5h 07m 50.9s	−05° 05′ 11″	2.79
γ	3h 58m 01.7s	−13° 30′ 31″	2.95
δ	3h 43m 14.8s	−09° 45′ 48″	3.54
ε	3h 32m 55.8s	−09° 27′ 30″	3.73
ζ	3h 15m 49.9s	−08° 49′ 11″	4.80
η	2h 56m 25.6s	−08° 53′ 53″	3.89
θ$_1$	2h 58m 15.6s	−40° 18′ 17″	3.42
θ$_2$	2h 58m 16.2s	−40° 18′ 16″	4.42
ι	2h 40m 40.0s	−39° 51′ 19″	4.11
κ	2h 26m 59.1s	−47° 42′ 14″	4.25
λ	5h 09m 08.7s	−08° 45′ 15″	4.27
μ	4h 45m 30.1s	−03° 15′ 17″	4.02
ν	4h 36m 19.1s	−03° 21′ 09″	3.93
ξ	4h 23m 40.8s	−03° 44′ 44″	5.17
o$_1$	4h 11m 51.9s	−06° 50′ 16″	4.04
o$_2$	4h 15m 16.3s	−07° 39′ 10″	4.43
π	3h 46m 08.4s	−12° 06′ 06″	4.42
ρ$_1$	3h 01m 09.9s	−07° 39′ 47″	5.75
ρ$_2$	3h 02m 42.2s	−07° 41′ 07″	5.32
ρ$_3$	3h 04m 16.3s	−07° 36′ 03″	5.26
τ$_1$	2h 45m 06.1s	−18° 34′ 21″	4.47
τ$_2$	2h 51m 02.2s	−21° 00′ 15″	4.75
τ$_3$	3h 02m 23.5s	−23° 37′ 28″	4.09
τ$_4$	3h 19m 30.9s	−21° 45′ 28″	3.69
τ$_5$	3h 33m 47.2s	−21° 37′ 59″	4.27
τ$_6$	3h 46m 50.8s	−23° 14′ 59″	4.23
τ$_7$	3h 47m 39.6s	−23° 52′ 29″	5.24

Star	Right ascension	Declination	Mag.
τ_8	3h 53m 42.6s	−24° 36′ 45″	4.65
τ_9	3h 59m 55.4s	−24° 00′ 59″	4.66
υ_1	4h 17m 53.6s	−33° 47′ 54″	3.56
υ_2	4h 33m 30.6s	−29° 46′ 00″	4.51
υ_3	4h 35m 33.0s	−30° 33′ 45″	3.82
φ	2h 16m 30.6s	−51° 30′ 44″	3.56
χ	1h 55m 57.4s	−51° 36′ 32″	3.70
ψ	5h 01m 26.3s	−07° 10′ 26″	4.81
ω	4h 52m 53.6s	−05° 27′ 10″	4.39

Deep sky object	Description	Right ascension	Declination	Mag.
NGC 1232	Spiral Galaxy	03h 09m 45.1s	−20° 34.7′	9.9
NGC 1291	Galaxy	03h 17m 18.3s	−41° 06.4′	9.4
NGC 1407	Galaxy	03h 40m 11.8s	−18° 34.8′	9.8
NGC 1535	Planetary Nebula	04h 14m 15.8s	−12° 44.4′	9.6

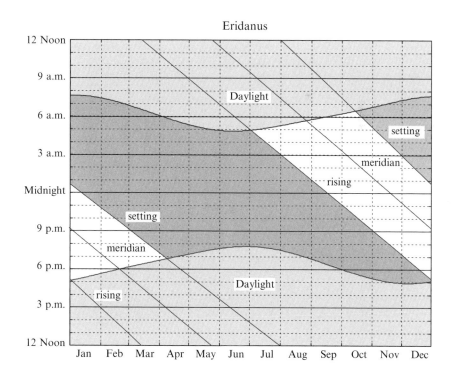

Eridanus

Fornax

the Furnace

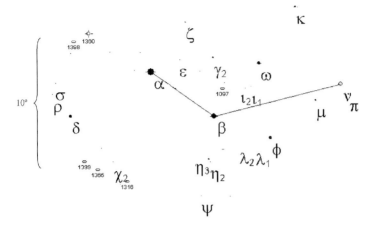

Star	Right ascension	Declination	Mag.
α	3h 12m 04.2s	−28° 59′ 14″	3.87
β	2h 49m 05.4s	−32° 24′ 22″	4.46
γ₁	2h 49m 50.9s	−24° 33′ 37″	6.14
γ₂	2h 49m 54.1s	−27° 56′ 31″	5.39
δ	3h 42m 14.9s	−31° 56′ 18″	5.00
ε	3h 01m 37.6s	−28° 05′ 29″	5.89
ζ	2h 59m 36.1s	−25° 16′ 27″	5.71
η₁	2h 47m 33.6s	−35° 33′ 03″	6.51
η₂	2h 50m 14.7s	−35° 50′ 37″	5.92
η₃	2h 50m 40.3s	−35° 40′ 34″	5.47
ι₁	2h 36m 09.2s	−30° 02′ 41″	5.75
ι₂	2h 38m 18.6s	−30° 11′ 40″	5.83
κ	2h 22m 32.5s	−23° 48′ 59″	5.20
λ₁	2h 33m 07.0s	−34° 39′ 00″	5.90
λ₂	2h 36m 58.5s	−34° 34′ 42″	5.79
μ	2h 12m 54.4s	−30° 43′ 26″	5.28
ν	2h 04m 29.4s	−29° 17′ 49″	4.69
π	2h 01m 14.7s	−30° 00′ 07″	5.35
ρ	3h 47m 56.0s	−30° 10′ 06″	5.54
σ	3h 46m 27.3s	−29° 20′ 17″	5.90
τ	3h 38m 47.6s	−27° 56′ 35″	6.01
φ	2h 28m 01.6s	−33° 48′ 40″	5.14
χ₁	3h 25m 55.7s	−35° 55′ 15″	6.39
χ₂	3h 27m 33.3s	−35° 40′ 54″	5.71
χ₃	3h 28m 11.4s	−35° 51′ 12″	6.50
ψ	2h 53m 34.2s	−38° 26′ 13″	5.92
ω	2h 33m 50.6s	−28° 13′ 57″	4.90

Deep sky object	Description	Right ascension	Declination	Mag.
NGC 1097	Galaxy	02h 46m 19.0s	−30° 16.5′	9.3
NGC 1316	Galaxy	03h 22m 41.7s	−37° 12.5′	8.8
NGC 1360	Planetary Nebula	03h 33m 14.6s	−25° 52.3′	9.4
NGC 1365	Galaxy	03h 33m 36.3s	−36° 08.4′	9.5
NGC 1398	Galaxy	03h 38m 52.0s	−26° 20.2′	9.7
NGC 1399	Galaxy	03h 38m 29.0s	−35° 27.1′	9.8

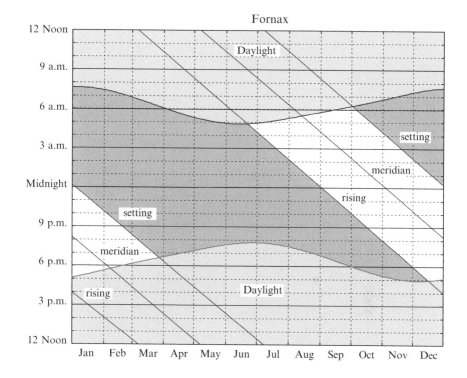

Fornax

Gemini

the Twins

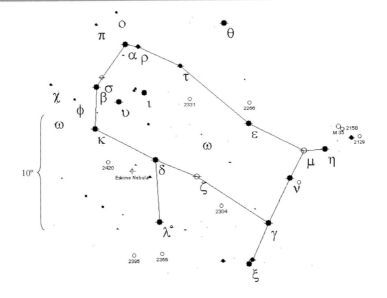

Star	Right ascension	Declination	Mag.
α_1	7h 34m 35.9s	+31° 53′ 18″	1.58
α_2	7h 34m 35.9s	+31° 53′ 18″	1.59
β	7h 45m 18.9s	+28° 01′ 34″	1.14
γ	6h 37m 42.7s	+16° 23′ 57″	1.93
δ	7h 20m 07.3s	+21° 58′ 56″	3.53
ε	6h 43m 55.9s	+25° 07′ 52″	2.98
ζ	7h 04m 06.5s	+20° 34′ 13″	3.79
η	6h 14m 52.6s	+22° 30′ 24″	3.28
θ	6h 52m 47.3s	+33° 57′ 40″	3.60
ι	7h 25m 43.5s	+27° 47′ 53″	3.79
κ	7h 44m 26.8s	+24° 23′ 53″	3.57
λ	7h 18m 05.5s	+16° 32′ 25″	3.58
μ	6h 22m 57.6s	+22° 30′ 49″	2.88
ν	6h 28m 57.7s	+20° 12′ 43″	4.15
ξ	6h 45m 17.3s	+12° 53′ 44″	3.36
ο	7h 39m 09.9s	+34° 35′ 04″	4.90
π	7h 47m 30.3s	+33° 24′ 56″	5.14
ρ	7h 29m 06.6s	+31° 47′ 04″	4.18
σ	7h 43m 18.7s	+28° 53′ 00″	4.28
τ	7h 11m 08.3s	+30° 14′ 43″	4.41
υ	7h 35m 55.3s	+26° 53′ 44″	4.06
φ	7h 53m 29.7s	+26° 45′ 57″	4.97
χ	8h 03m 31.0s	+27° 47′ 39″	4.94
ω	7h 02m 24.7s	+24° 12′ 56″	5.18

Deep sky object	Description	Right ascension	Declination	Mag.
M35	Open Cluster	06h 08m 55.9s	+24° 21.5′	5.1
NGC 2129	Open Cluster	06h 01m 06.5s	+23° 19.3′	6.7
NGC 2158	Open Cluster	06h 07m 25.6s	+24° 05.8′	8.6
NGC 2266	Open Cluster	06h 43m 19.2s	+26° 58.2′	9.5
NGC 2304	Open Cluster	06h 55m 11.9s	+17° 59.3′	10.0
NGC 2331	Open Cluster	07h 06m 59.8s	+27° 15.7′	8.5
NGC 2355	Open Cluster	07h 16m 59.3s	+13° 45.0′	9.7
NGC 2392	The Eskimo Nebula	07h 29m 10.8s	+20° 54.7′	9.2
NGC 2395	Open Cluster	07h 27m 12.9s	+13° 36.5′	8.0
NGC 2420	Open Cluster	07h 38m 23.9s	+21° 34.5′	8.3

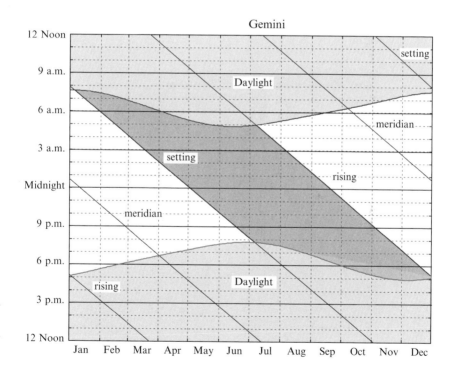

Gemini

Grus

the Crane

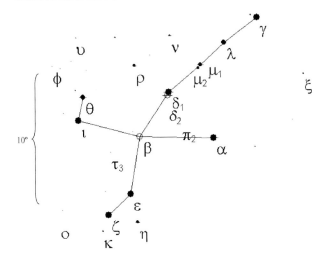

Star	Right ascension	Declination	Mag.
α	22h 08m 13.9s	−46° 57′ 40″	1.74
β	22h 42m 40.0s	−46° 53′ 05″	2.10
γ	21h 53m 55.6s	−37° 21′ 54″	3.01
δ	22h 29m 16.1s	−43° 29′ 45″	3.97
δ	22h 29m 45.4s	−43° 44′ 58″	4.11
ε	22h 48m 33.2s	−51° 19′ 01″	3.49
ζ	23h 00m 52.8s	−52° 45′ 15″	4.12
η	22h 45m 37.8s	−53° 30′ 00″	4.85
θ	23h 06m 52.7s	−43° 31′ 14″	4.28
ι	23h 10m 21.5s	−45° 14′ 48″	3.90
κ	23h 04m 39.5s	−53° 57′ 55″	5.37
λ	22h 06m 06.8s	−39° 32′ 36″	4.46
μ	22h 15m 36.9s	−41° 20′ 48″	4.79
μ	22h 16m 26.5s	−41° 37′ 39″	5.10
ν	22h 28m 39.1s	−39° 07′ 55″	5.47
ξ	21h 32m 05.7s	−41° 10′ 46″	5.29
ο	23h 26m 36.4s	−52° 43′ 18″	5.52
π	22h 22m 43.8s	−45° 56′ 52″	6.62
π	22h 23m 07.9s	−45° 55′ 42″	5.62
ρ	22h 43m 30.0s	−41° 24′ 52″	4.85
σ	22h 36m 29.2s	−40° 34′ 58″	6.28
σ	22h 36m 58.7s	−40° 35′ 28″	5.86
τ	22h 53m 37.8s	−48° 35′ 53″	6.04
τ	22h 56m 47.7s	−47° 58′ 09″	5.70
υ	23h 06m 53.5s	−38° 53′ 32″	5.61
φ	23h 18m 09.8s	−40° 49′ 29″	5.53

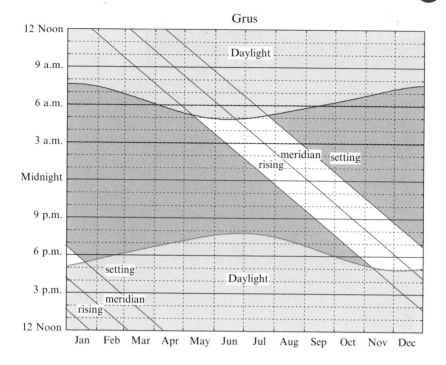

Hercules

the Warrior

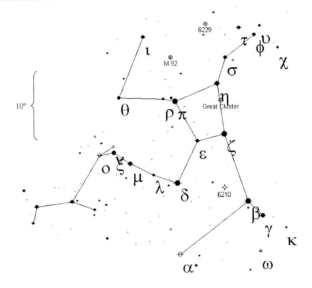

Star	Right ascension	Declination	Mag.
α_1	17h 14m 38.8s	+14° 23′ 25″	3.48
α_2	17h 14m 39.1s	+14° 23′ 24″	5.39
β	16h 30m 13.1s	+21° 29′ 22″	2.77
γ	16h 21m 55.1s	+19° 09′ 11″	3.75
δ	17h 15m 01.8s	+24° 50′ 21″	3.14
ε	17h 00m 17.3s	+30° 55′ 35″	3.92
ζ	16h 41m 17.1s	+31° 36′ 10″	2.81
η	16h 42m 53.7s	+38° 55′ 20″	3.53
θ	17h 56m 15.1s	+37° 15′ 02″	3.86
ι	17h 39m 27.8s	+46° 00′ 23″	3.80
κ_1	16h 08m 04.4s	+17° 02′ 49″	5.00
κ_2	16h 08m 04.8s	+17° 03′ 16″	6.25
λ	17h 30m 44.2s	+26° 06′ 38″	4.41
μ	17h 46m 27.5s	+27° 43′ 15″	3.42
ν	17h 58m 30.1s	+30° 11′ 22″	4.41
ξ	17h 57m 45.8s	+29° 14′ 52″	3.70
o	18h 07m 32.5s	+28° 45′ 45″	3.83
π	17h 15m 02.7s	+36° 48′ 33″	3.16
ρ_1	17h 23m 40.6s	+37° 08′ 48″	5.47
ρ_2	17h 23m 40.9s	+37° 08′ 45″	4.52
σ	16h 34m 06.1s	+42° 26′ 13″	4.20
τ	16h 19m 44.3s	+46° 18′ 48″	3.89
υ	16h 02m 47.8s	+46° 02′ 12″	4.76
φ	16h 08m 46.1s	+44° 56′ 06″	4.26
χ	15h 52m 40.4s	+42° 27′ 06″	4.62
ω	16h 25m 24.9s	+14° 02′ 00″	4.57

Deep sky object	Description	Right ascension	Declination	Mag.
M13	The Great Cluster	16h 41m 42.0s	+36° 28.0′	5.7
M92	Globular Cluster	17h 17m 06.0s	+43° 08.0′	6.4
NGC 6210	Planetary Nebula	16h 44m 29.5s	+23° 48.0′	8.8
NGC 6229	Globular Cluster	16h 47m 00.0s	+47° 32.0′	9.4

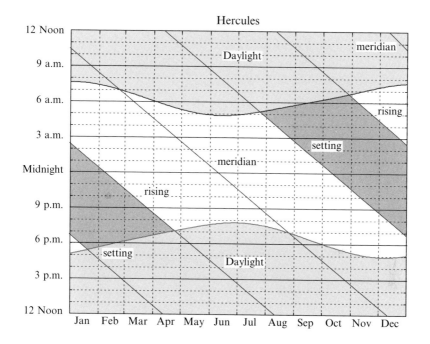

Horologium

the Clock

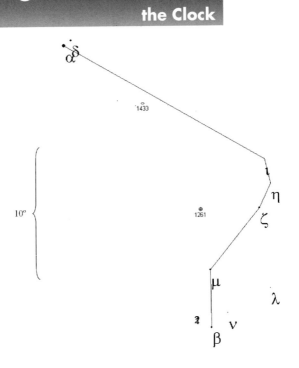

Star	Right ascension	Declination	Mag.
α	04h 14m 00.1s	−42° 17′ 40″	3.86
β	02h 58m 47.8s	−64° 04′ 16″	4.99
γ	02h 45m 27.5s	−63° 42′ 16″	5.74
δ	04h 10m 50.5s	−41° 59′ 37″	4.93
ζ	02h 40m 39.6s	−54° 33′ 00″	5.21
η	02h 37m 24.2s	−52° 32′ 36″	5.31
ι	02h 42m 33.4s	−50° 48′ 01″	5.41
λ	02h 24m 53.9s	−60° 18′ 43″	5.35
μ	03h 03m 36.8s	−59° 44′ 16″	5.11
ν	02h 49m 01.5s	−62° 48′ 24″	5.26

Deep sky object	Description	Right ascension	Declination	Mag.
NGC 1261	Globular Cluster	03h 12m 15.3s	−55° 13.0′	8.4
NGC 1433	Galaxy	03h 42m 01.4s	−47° 13.3′	10.0

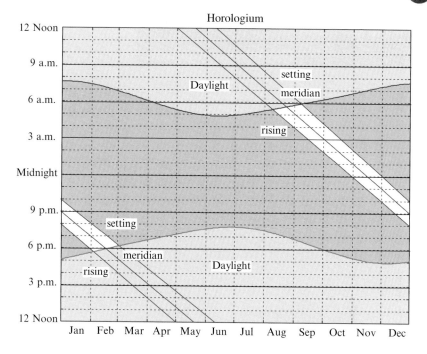

Horologium

Hydra

the Water Serpent

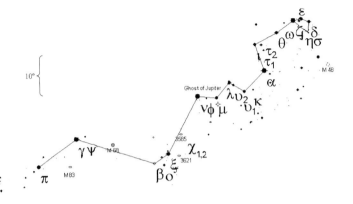

Star	Right ascension	Declination	Mag.
α	09h 27m 35.2s	−08° 39′ 31″	1.98
β	11h 52m 54.5s	−33° 54′ 28″	4.28
γ	13h 18m 55.2s	−23° 10′ 18″	3.00
δ	08h 37m 39.3s	+05° 42′ 13″	4.16
ε	08h 46m 46.5s	+06° 25′ 08″	3.38
ζ	08h 55m 23.6s	+05° 56′ 44″	3.11
η	08h 43m 13.4s	+03° 23′ 55″	4.30
θ	09h 14m 21.8s	+02° 18′ 51″	3.88
ι	09h 39m 51.3s	−01° 08′ 34″	3.91
κ	09h 40m 18.3s	−14° 19′ 56″	5.06
λ	10h 10m 35.2s	−12° 21′ 15″	3.61
μ	10h 26m 05.4s	−16° 50′ 11″	3.81
ν	10h 49m 37.4s	−16° 11′ 37″	3.11
ξ	11h 33m 00.1s	−31° 51′ 27″	3.54
o	11h 40m 12.7s	−34° 44′ 41″	4.70
π	14h 06m 22.2s	−26° 40′ 56″	3.27
ρ	08h 48m 25.9s	+05° 50′ 16″	4.36
σ	08h 38m 45.4s	+03° 20′ 29″	4.44
τ_1	09h 29m 08.8s	−02° 46′ 08″	4.60
τ_2	09h 31m 58.9s	−01° 11′ 05″	4.57
υ_1	09h 51m 28.6s	−14° 50′ 48″	4.12
υ_2	10h 05m 07.4s	−13° 03′ 53″	4.60
φ_1	10h 36m 16.6s	−16° 20′ 40″	6.03
φ_2	10h 38m 34.9s	−16° 52′ 36″	4.91
χ_1	11h 05m 19.9s	−27° 17′ 36″	4.94
χ_2	11h 05m 57.5s	−27° 17′ 16″	5.71
ψ	13h 09m 03.2s	−23° 07′ 05″	4.95
ω	09h 05m 58.3s	+05° 05′ 32″	4.97

Deep sky object	Description	Right ascension	Declination	Mag.
M48	Open Cluster	08h 13m 43.1s	–05° 45.0′	5.8
M68	Globular Cluster	12h 39m 28.0s	–26° 44.6′	7.7
M83	Galaxy	13h 37m 00.0s	–29° 52.0′	7.5
NGC 3242	Ghost of Jupiter Nebula	10h 24m 46.2s	–18° 38.6′	7.8
NGC 3585	Galaxy	11h 13m 17.1s	–26° 45.3′	10.0
NGC 3621	Galaxy	11h 18m 18.0s	–32° 49.0′	10.0

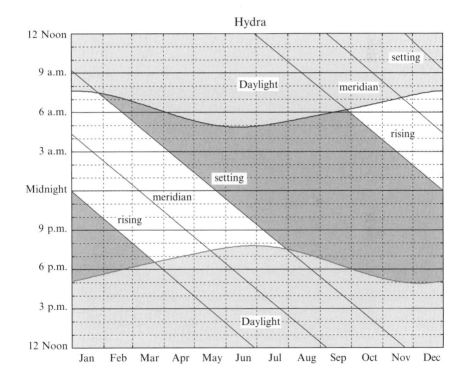

Hydra

Hydrus

the Sea Serpent

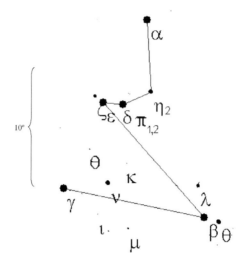

Star	Right ascension	Declination	Mag.
α	01h 58m 46.2s	−61° 34′ 12″	2.86
β	00h 25m 45.3s	−77° 15′ 16″	2.80
γ	03h 47m 14.5s	−74° 14′ 21″	3.24
δ	02h 21m 45.1s	−68° 39′ 34″	4.09
ε	02h 39m 35.5s	−68° 16′ 01″	4.11
ζ	02h 45m 32.6s	−67° 37′ 00″	4.84
η	01h 54m 56.1s	−67° 38′ 51″	4.69
θ	03h 02m 15.5s	−71° 54′ 09″	5.53
ι	03h 15m 57.7s	−77° 23′ 19″	5.52
κ	02h 22m 52.3s	−73° 38′ 45″	5.01
λ	00h 48m 35.4s	−74° 55′ 25″	5.07
μ	02h 31m 40.7s	−79° 06′ 34″	5.28
ν	02h 50m 28.7s	−75° 04′ 01″	4.75
π_1	02h 14m 14.8s	−67° 50′ 31″	5.55
π_2	02h 15m 28.6s	−67° 44′ 48″	5.69
σ	01h 55m 50.4s	−78° 20′ 55″	6.16
τ_1	01h 41m 21.4s	−79° 08′ 54″	6.33
τ_2	01h 47m 46.7s	−80° 10′ 36″	6.06

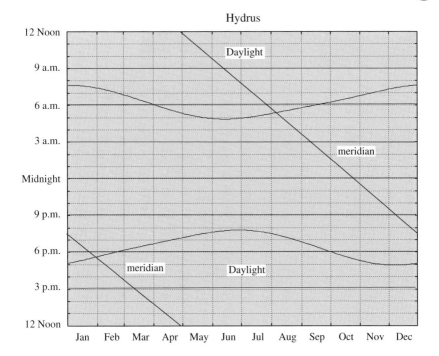

Hydrus

The constellation Hydrus is not visible from mid-northern latitudes

Indus
the American Indian

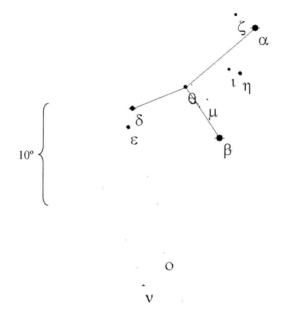

Star	Right ascension	Declination	Mag.
α	20h 37m 34.0s	−47° 17′ 29″	3.11
β	20h 54m 48.5s	−58° 27′ 15″	3.65
γ	21h 26m 15.4s	−54° 39′ 38″	6.12
δ	21h 57m 55.0s	−54° 59′ 34″	4.40
ε	22h 03m 21.3s	−56° 47′ 10″	4.69
ζ	20h 49m 28.9s	−46° 13′ 37″	4.89
η	20h 44m 02.2s	−51° 55′ 16″	4.51
θ	21h 19m 51.9s	−53° 26′ 59″	4.39
ι	20h 51m 30.0s	−51° 36′ 30″	5.05
κ₁	21h 58m 29.9s	−59° 00′ 45″	6.12
κ₂	22h 05m 50.9s	−59° 38′ 10″	5.62
μ	21h 05m 14.1s	−54° 43′ 37″	5.16
ν	22h 24m 36.7s	−72° 15′ 20″	5.29
ο	21h 50m 47.2s	−69° 37′ 47″	5.53
π	21h 56m 13.9s	−57° 53′ 59″	6.19
ρ	22h 54m 39.5s	−70° 04′ 26″	6.05

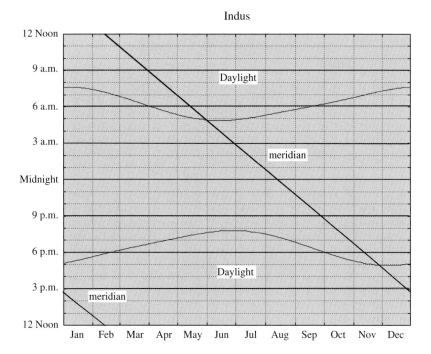

Indus

The constellation Indus is not visible from mid-northern latitudes

Lacerta
the Lizard

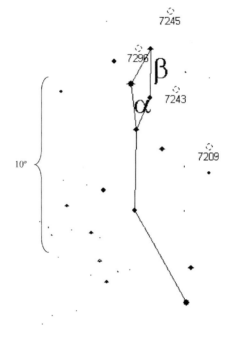

Star	Right ascension	Declination	Mag.
α	22h 31m 17.4s	+50° 16′ 57″	3.77
β	22h 23m 33.5s	+52° 13′ 45″	4.43

Deep sky object	Description	Right ascension	Declination	Mag.
NGC 7209	Open Cluster	22h 05m 12.0s	+46° 30.0′	7.7
NGC 7243	Open Cluster	22h 15m 18.0s	+49° 53.0′	6.4
NGC 7245	Open Cluster	22h 15m 18.0s	+54° 20.0′	9.2
NGC 7296	Open Cluster	22h 28m 12.0s	+52° 17.0′	9.7

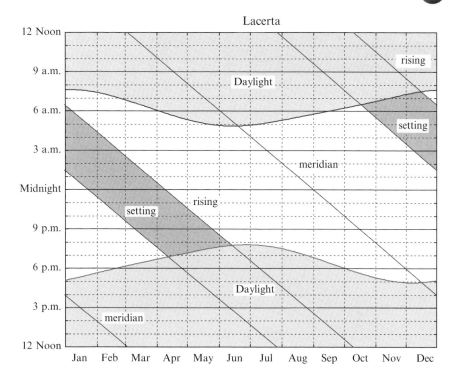

Lacerta

Leo
the Lion

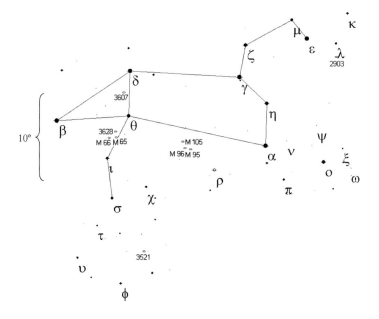

Star	Right ascension	Declination	Mag.
α	10h 08m 22.3s	+11° 58′ 02″	1.35
β	11h 49m 03.5s	+14° 34′ 19″	2.14
γ₁	10h 19m 58.3s	+19° 50′ 30″	2.61
γ₂	10h 19m 58.6s	+19° 50′ 25″	3.80
δ	11h 14m 06.4s	+20° 31′ 25″	2.56
ε	09h 45m 51.0s	+23° 46′ 27″	2.98
ζ	10h 16m 41.3s	+23° 25′ 02″	3.44
η	10h 07m 19.9s	+16° 45′ 45″	3.52
θ	11h 14m 14.3s	+15° 25′ 46″	3.34
ι	11h 23m 55.4s	+10° 31′ 45″	3.94
κ	09h 24m 39.2s	+26° 10′ 56″	4.46
λ	09h 31m 43.1s	+22° 58′ 04″	4.31
μ	09h 52m 45.8s	+26° 00′ 25″	3.88
ν	09h 58m 13.3s	+12° 26′ 41″	5.26
ξ	09h 31m 56.7s	+11° 17′ 59″	4.97
ο	09h 41m 09.0s	+09° 53′ 32″	3.52
π	10h 00m 12.7s	+08° 02′ 39″	4.70
ρ	10h 32m 48.6s	+09° 18′ 24″	3.85
σ	11h 21m 08.1s	+06° 01′ 46″	4.05
τ	11h 27m 56.2s	+02° 51′ 22″	4.95
υ	11h 36m 56.9s	−00° 49′ 26″	4.30
φ	11h 16m 39.6s	−03° 39′ 06″	4.47
χ	11h 05m 01.0s	+07° 20′ 10″	4.63
ψ	09h 43m 43.8s	+14° 01′ 18″	5.35
ω	09h 28m 27.4s	+09° 03′ 24″	5.41

Deep sky object	Description	Right ascension	Declination	Mag.
M65	Galaxy	11h 18m 54.0s	+13° 05.0′	9.3
M66	Galaxy	11h 20m 12.0s	+12° 59.0′	9.0
M95	Galaxy	10h 43m 57.7s	+11° 42.2′	9.7
M96	Galaxy	10h 46m 45.9s	+11° 49.4′	9.2
M105	Galaxy	10h 47m 48.0s	+12° 35.0′	9.3
NGC 2903	Spiral Galaxy	09h 32m 09.9s	+21° 30.1′	8.9
NGC 3521	Galaxy	11h 05m 48.0s	−00° 02.0′	8.9
NGC 3607	Galaxy	11h 16m 54.0s	+18° 03.0′	10.0
NGC 3628	Galaxy	11h 20m 18.0s	+13° 36.0′	9.5

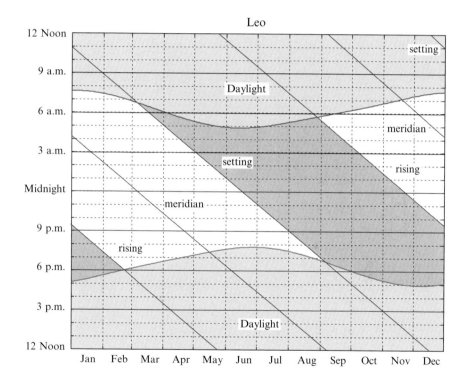

Leo Minor
the Smaller Lion

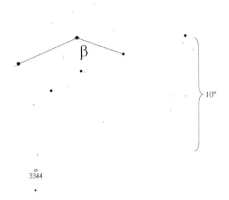

Star	Right ascension	Declination	Mag.
β	10h 27m 52.9s	+36° 42′ 26″	4.21

Deep sky object	Description	Right ascension	Declination	Mag.
NGC 3344	Galaxy	10h 43m 30.7s	+24° 55.3′	10.0

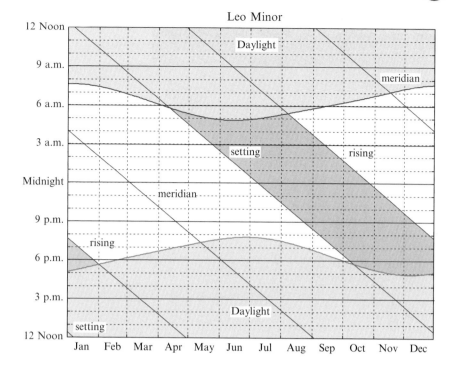
Leo Minor

Lepus

the Hare

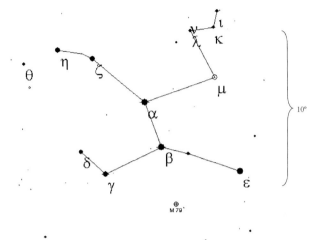

Star	Right ascension	Declination	Mag.
α	5h 32m 43.7s	−17° 49′ 20″	2.58
β	5h 28m 14.7s	−20° 45′ 34″	2.84
γ	5h 44m 27.8s	−22° 26′ 54″	3.60
δ	5h 51m 19.2s	−20° 52′ 45″	3.81
ε	5h 05m 27.6s	−22° 22′ 16″	3.19
ζ	5h 46m 57.3s	−14° 49′ 19″	3.55
η	5h 56m 24.2s	−14° 10′ 04″	3.71
θ	6h 06m 09.3s	−14° 56′ 07″	4.67
ι	5h 12m 17.8s	−11° 52′ 09″	4.45
κ	5h 13m 13.8s	−12° 56′ 30″	4.36
λ	5h 19m 34.4s	−13° 10′ 37″	4.29
μ	5h 12m 55.8s	−16° 12′ 20″	3.31
ν	5h 19m 59.0s	−12° 18′ 56″	5.30

Deep sky object	Description	Right ascension	Declination	Mag.
M79	Globular Cluster	05h 24m 10.6s	−24° 31.5′	7.8

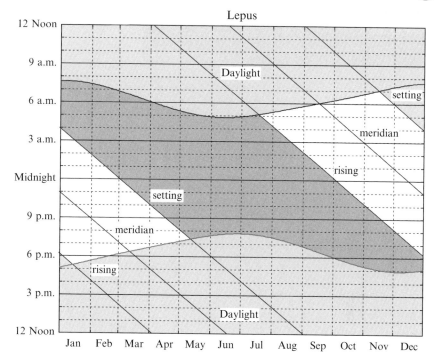

Lepus

Libra

the Scales

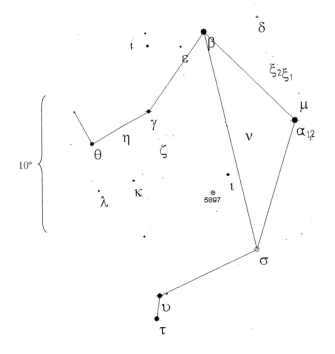

Star	Right ascension	Declination	Mag.
α_1	14h 50m 41.1s	−15° 59′ 50″	5.15
α_2	14h 50m 52.6s	−16° 02′ 31″	2.75
β	15h 17m 00.3s	−09° 22′ 59″	2.61
γ	15h 35m 31.5s	−14° 47′ 22″	3.91
δ	15h 00m 58.3s	−08° 31′ 08″	4.92
ε	15h 24m 11.8s	−10° 19′ 20″	4.94
ζ_1	15h 28m 15.3s	−16° 42′ 59″	5.64
ζ_2	15h 30m 40.3s	−16° 36′ 34″	5.82
ζ_3	15h 32m 55.1s	−16° 51′ 11″	5.50
η	15h 44m 04.3s	−15° 40′ 22″	5.41
θ	15h 53m 49.4s	−16° 43′ 46″	4.15
ι_1	15h 12m 13.2s	−19° 47′ 30″	4.54
ι_2	15h 13m 19.1s	−19° 38′ 51″	6.08
κ	15h 41m 56.7s	−19° 40′ 44″	4.74
λ	15h 53m 20.0s	−20° 10′ 02″	5.03
μ	14h 49m 19.0s	−14° 08′ 56″	5.31
ν	15h 06m 37.5s	−16° 15′ 24″	5.20
ξ_1	14h 54m 22.8s	−11° 53′ 54″	5.80
ξ_2	14h 56m 46.0s	−11° 24′ 35″	5.46
o_1	15h 21m 01.2s	−15° 32′ 54″	6.30
σ_2	15h 04m 04.1s	−25° 16′ 55″	3.29
τ	15h 38m 39.3s	−29° 46′ 40″	3.66
υ	15h 37m 01.4s	−28° 08′ 06″	3.58

Deep sky object	Description	Right ascension	Declination	Mag.
NGC 5897	Globular Cluster	15h 17m 24.0s	−21° 01.0′	8.6

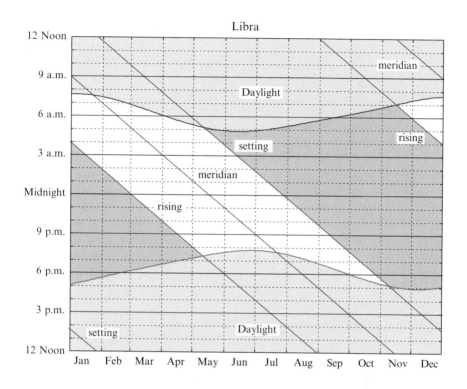

Libra

Lupus

the Wolf

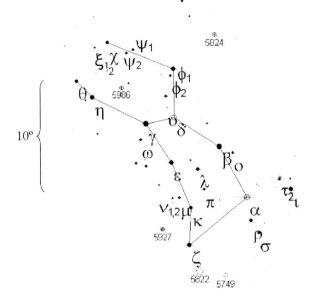

Star	Right ascension	Declination	Mag.
α	14h 41m 55.7s	−47° 23′ 17″	2.30
β	14h 58m 31.8s	−43° 08′ 02″	2.68
γ	15h 35m 08.3s	−41° 10′ 00″	2.78
δ	15h 21m 22.2s	−40° 38′ 52″	3.22
ε	15h 22m 40.8s	−44° 41′ 22″	3.37
ζ	15h 12m 17.0s	−52° 05′ 57″	3.41
η	16h 00m 07.2s	−38° 23′ 49″	3.41
θ	16h 06m 35.4s	−36° 48′ 08″	4.23
ι	14h 19m 24.1s	−46° 03′ 28″	3.55
κ₁	15h 11m 56.0s	−48° 44′ 16″	3.87
κ₂	15h 11m 57.5s	−48° 44′ 37″	5.69
λ	15h 08m 50.5s	−45° 16′ 47″	4.05
μ	15h 18m 31.9s	−47° 52′ 30″	4.27
ν₁	15h 22m 08.2s	−47° 55′ 40″	5.00
ν₂	15h 21m 48.1s	−48° 19′ 03″	5.65
ξ₁	15h 56m 53.4s	−33° 57′ 59″	5.37
ξ₂	15h 56m 54.1s	−33° 57′ 51″	5.73
o	14h 51m 38.3s	−43° 34′ 31″	4.32
π₁	15h 05m 07.1s	−47° 03′ 04″	4.72
π₂	15h 05m 07.1s	−47° 03′ 04″	4.82
ρ	14h 37m 53.1s	−49° 25′ 32″	4.05
σ	14h 32m 36.8s	−50° 27′ 25″	4.42
τ₁	14h 26m 08.1s	−45° 13′ 17″	4.56
τ₂	14h 26m 10.7s	−45° 22′ 46″	4.35
υ	15h 24m 44.8s	−39° 42′ 37″	5.37

Star	Right ascension	Declination	Mag.
φ_1	15h 21m 48.3s	–36° 15′ 41″	3.56
φ_2	15h 23m 09.2s	–36° 51′ 30″	4.54
χ	15h 50m 57.4s	–33° 37′ 38″	3.95
ψ_1	15h 39m 45.9s	–34° 24′ 42″	4.67
ψ_2	15h 42m 40.9s	–34° 42′ 38″	4.75
ω	15h 38m 03.1s	–42° 34′ 02″	4.33

Deep sky object	Description	Right ascension	Declination	Mag.
NGC 5749	Open Cluster	14h 48m 54.0s	–54° 31.0′	8.8
NGC 5822	Open Cluster	15h 05m 12.0s	–54° 21.0′	6.5
NGC 5824	Globular Cluster	15h 04m 00.0s	–33° 04.0′	7.8
NGC 5927	Globular Cluster	15h 28m 00.0s	–50° 40.0′	8.3
NGC 5986	Globular Cluster	15h 46m 06.0s	–37° 47.0′	7.5

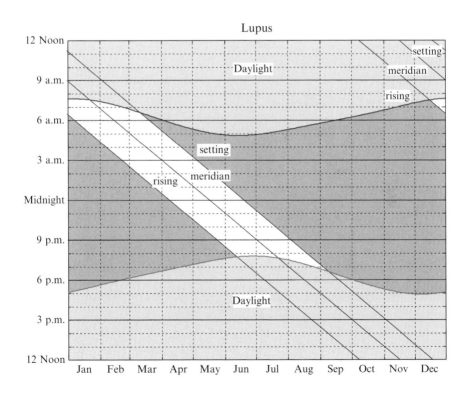

Lupus

Lynx

the Lynx

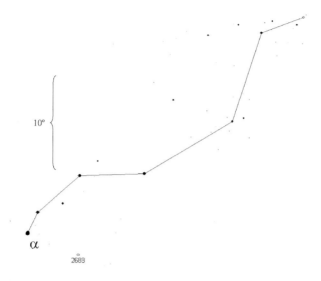

Star	Right ascension	Declination	Mag.
α	9h 21m 03.2s	+34° 23′ 33″	3.13

Deep sky object	Description	Right ascension	Declination	Mag.
NGC 2683	UFO Galaxy	08h 52m 41.8s	+33° 25.3′	9.7

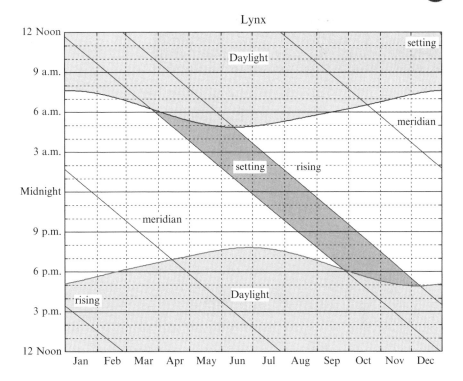

Lynx

Lyra

the Lyre

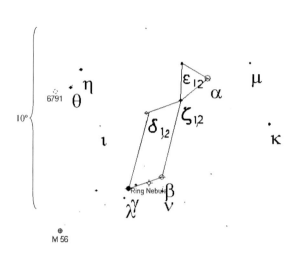

Star	Right ascension	Declination	Mag.
α	18h 36m 56.2s	+38° 47′ 01″	0.03
β	18h 50m 04.7s	+33° 21′ 46″	3.45
γ	18h 58m 56.5s	+32° 41′ 22″	3.24
δ_1	18h 53m 43.5s	+36° 58′ 18″	5.58
δ_2	18h 54m 30.1s	+36° 53′ 56″	4.30
ε_1	18h 44m 20.3s	+39° 40′ 12″	5.06
ε_2	18h 44m 20.2s	+39° 40′ 15″	6.02
ε_3	18h 44m 22.8s	+39° 36′ 46″	5.14
ε_4	18h 44m 22.8s	+39° 36′ 46″	5.37
ζ_1	18h 44m 46.3s	+37° 36′ 18″	4.36
ζ_2	18h 44m 48.1s	+37° 35′ 40″	5.73
η	19h 13m 45.4s	+39° 08′ 46″	4.39
θ	19h 16m 22.0s	+38° 08′ 01″	4.36
ι	19h 07m 18.0s	+36° 06′ 01″	5.28
κ	18h 19m 51.6s	+36° 03′ 52″	4.33
λ	19h 00m 00.8s	+32° 08′ 44″	4.93
μ	18h 24m 13.7s	+39° 30′ 26″	5.12
ν_1	18h 49m 45.8s	+32° 48′ 46″	5.91
ν_2	18h 49m 52.8s	+32° 33′ 03″	5.25

Deep sky object	Description	Right ascension	Declination	Mag.
M56	Globular Cluster	19h 16m 36.0s	+30° 11.0′	8.3
M57	The Ring Nebula	18h 53m 35.1s	+33° 01.7′	8.8
NGC 6791	Open Cluster	19h 20m 42.0s	+37° 51.0′	9.5

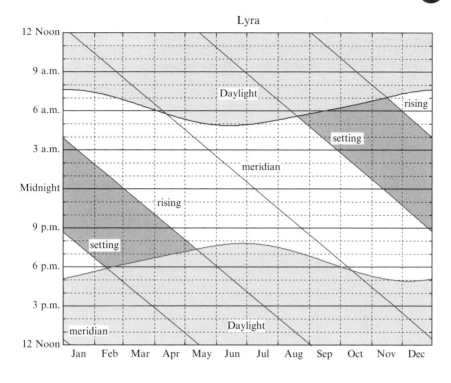

Mensa
the Table Mountain

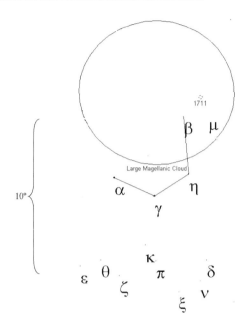

Star	Right ascension	Declination	Mag.
α	06h 10m 14.6s	−74° 45′ 11″	5.09
β	05h 02m 43.2s	−71° 18′ 50″	5.31
γ	05h 31m 53.0s	−76° 20′ 28″	5.19
δ	04h 17m 59.2s	−80° 12′ 51″	5.69
ε	07h 25m 38.2s	−79° 05′ 38″	5.53
ζ	06h 40m 02.8s	−80° 48′ 49″	5.64
η	04h 55m 11.3s	−74° 56′ 13″	5.47
θ	06h 56m 34.5s	−79° 25′ 13″	5.45
ι	05h 35m 36.5s	−78° 49′ 15″	6.05
κ	05h 50m 16.5s	−79° 21′ 41″	5.47
λ	05h 47m 48.9s	−72° 42′ 09″	6.53
μ	04h 43m 04.1s	−70° 55′ 52″	5.54
ν	04h 20m 58.0s	−81° 34′ 48″	5.79
ξ	04h 58m 50.9s	−82° 28′ 14″	5.85
π	05h 37m 08.8s	−80° 28′ 09″	5.65

Deep sky object	Description	Right ascension	Declination	Mag.
NGC 1711	Open Cluster	04h 50m 36.2s	−69° 59.1′	10.0

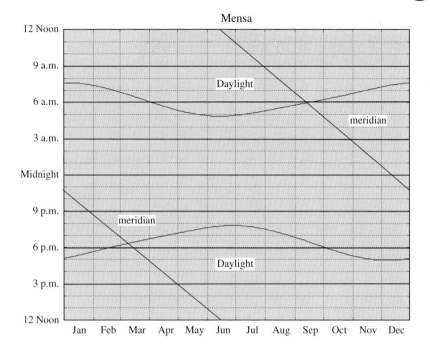

Mensa

The constellation Mensa is not visible from mid-northern latitudes

Microscopium
the Microscope

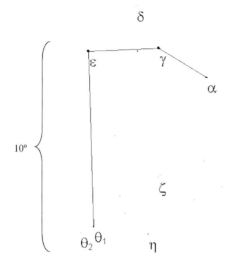

Star	Right ascension	Declination	Mag.
α	20h 49m 58.0s	−33° 46′ 48″	4.90
β	20h 51m 58.7s	−33° 10′ 38″	6.04
γ	21h 01m 17.4s	−32° 15′ 28″	4.67
δ	21h 06m 01.0s	−30° 07′ 31″	5.68
ε	21h 17m 56.2s	−32° 10′ 21″	4.71
ζ	21h 02m 57.8s	−38° 37′ 54″	5.30
η	21h 06m 25.4s	−41° 23′ 10″	5.53
θ_1	21h 20m 45.5s	−40° 48′ 35″	4.82
θ_2	21h 24m 24.7s	−41° 00′ 24″	5.77
ι	20h 48m 29.0s	−43° 59′ 19″	5.11
ν	20h 33m 55.0s	−44° 30′ 58″	5.11

Microscopium

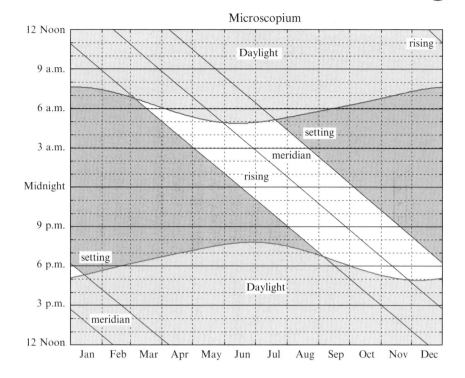

Monoceros

the Unicorn

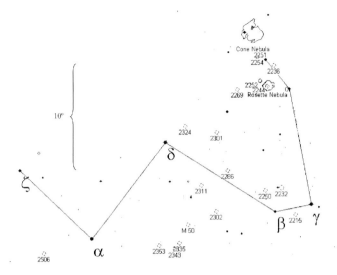

Star	Right ascension	Declination	Mag.
α	7h 41m 14.8s	–09° 33′ 04″	3.93
β₁	6h 28m 48.9s	–07° 01′ 59″	4.60
β₂	6h 28m 49.4s	–07° 02′ 04″	5.40
β₃	6h 28m 49.4s	–07° 02′ 04″	5.60
γ	6h 14m 51.3s	–06° 16′ 29″	3.98
δ	7h 11m 51.8s	–00° 29′ 34″	4.15
ε	6h 23m 46.0s	+04° 35′ 34″	4.44
ζ	8h 08m 35.6s	–02° 59′ 02″	4.34

Deep sky object	Description	Right ascension	Declination	Mag.
M50	Open Cluster	07h 02m 42.3s	−08° 23.5′	5.9
NGC 2215	Open Cluster	06h 20m 49.2s	−07° 17.0′	8.4
NGC 2232	Open Cluster	06h 28m 01.1s	−04° 50.8′	3.9
NGC 2236	Open Cluster	06h 29m 39.7s	+06° 49.8′	8.5
NGC 2237	Rosette Nebula	06h 30m 54.6s	+05° 02.9′	8.9
NGC 2239	Open Cluster	06h 31m 55.6s	+04° 56.6′	4.8
NGC 2250	Open Cluster	06h 34m 01.5s	−05° 00.3′	8.9
NGC 2251	Open Cluster	06h 34m 38.5s	+08° 22.0′	7.3
NGC 2252	Open Cluster	06h 34m 42.9s	+05° 22.0′	7.7
NGC 2254	Open Cluster	06h 35m 49.7s	+07° 40.4′	9.1
NGC 2264	Cone Nebula	06h 40m 58.3s	+09° 53.7′	3.9
	& Christmas Tree Cluster			
NGC 2269	Open Cluster	06h 43m 17.1s	+04° 37.5′	10.0
NGC 2286	Open Cluster	06h 47m 40.1s	−03° 08.9′	7.5
NGC 2301	Open Cluster	06h 51m 45.3s	+00° 27.5′	6.0
NGC 2302	Open Cluster	06h 51m 56.7s	−07° 05.1′	8.9
NGC 2311	Open Cluster	06h 57m 47.5s	−04° 36.7′	9.6
NGC 2324	Open Cluster	07h 04m 07.9s	+01° 02.7′	8.4
NGC 2335	Open Cluster	07h 06m 49.5s	−10° 01.7′	7.2
NGC 2343	Open Cluster	07h 08m 06.8s	−10° 37.0′	6.7
NGC 2353	Open Cluster	07h 14m 30.3s	−10° 15.9′	7.1
NGC 2506	Open Cluster	08h 00m 01.7s	−10° 46.2′	7.6

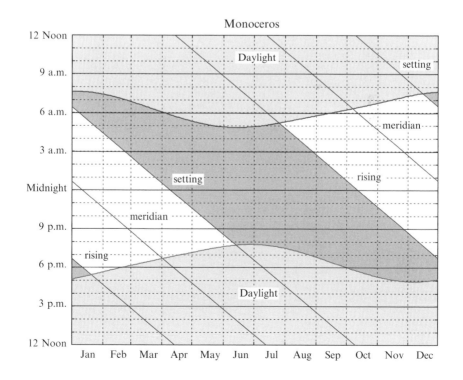

Musca

the Fly

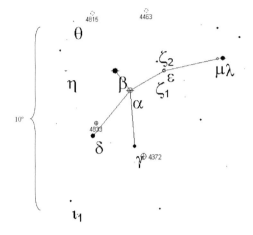

Star	Right ascension	Declination	Mag.
α	12h 37m 11.0s	−69° 08′ 08″	2.69
β	12h 46m 16.9s	−68° 06′ 29″	3.05
γ	12h 32m 28.0s	−72° 07′ 58″	3.87
δ	13h 02m 16.3s	−71° 32′ 56″	3.62
ε	12h 17m 34.2s	−67° 57′ 38″	4.11
ζ₁	12h 22m 07.4s	−67° 31′ 20″	5.15
ζ₂	12h 22m 11.9s	−68° 18′ 27″	5.74
η	13h 15m 14.9s	−67° 53′ 41″	4.80
θ	13h 08m 07.0s	−65° 18′ 22″	5.51
ι₁	13h 25m 07.2s	−74° 53′ 16″	5.05
ι₂	13h 27m 18.3s	−74° 41′ 30″	6.63
λ	11h 45m 36.4s	−66° 43′ 43″	3.64
μ	11h 48m 14.4s	−66° 48′ 53″	4.72

Deep sky object	Description	Right ascension	Declination	Mag.
NGC 4372	Globular Cluster	12h 25m 45.4s	−72° 39.5′	7.8
NGC 4463	Open Cluster	12h 29m 55.2s	−64° 47.4′	7.2
NGC 4815	Open Cluster	12h 58m 00.0s	−64° 57.0′	8.6
NGC 4833	Globular Cluster	12h 59m 36.0s	−70° 53.0′	7.4

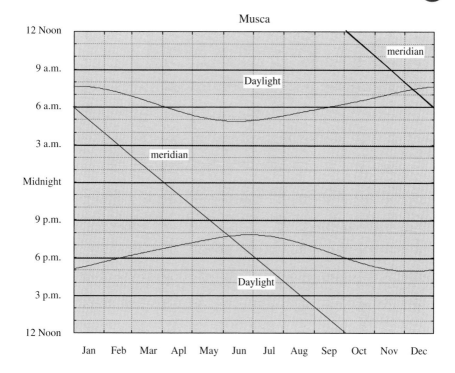

The constellation Musca is not visible from mid-northern latitudes

Norma
the Carpenter's Square

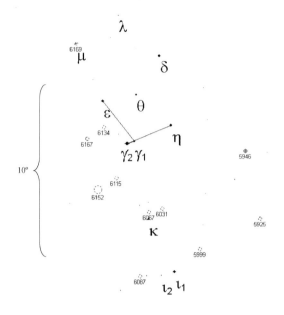

Star	Right ascension	Declination	Mag.
γ_1	16h 17m 00.8s	−50° 04′ 05″	4.99
γ_2	16h 19m 50.3s	−50° 09′ 20″	4.02
δ	16h 06m 29.3s	−45° 10′ 23″	4.72
ε	16h 27m 11.0s	−47° 33′ 18″	4.47
ζ	16h 13m 22.5s	−55° 32′ 27″	5.81
η	16h 03m 12.7s	−49° 13′ 47″	4.65
θ	16h 15m 15.2s	−47° 22′ 20″	5.14
ι_1	16h 03m 31.9s	−57° 46′ 31″	4.63
ι_2	16h 09m 18.4s	−57° 56′ 03″	5.57
κ	16h 13m 28.6s	−54° 37′ 50″	4.94
λ	16h 19m 17.6s	−42° 40′ 26″	5.45
μ	16h 34m 04.8s	−44° 02′ 43″	4.94

Deep sky object	Description	Right ascension	Declination	Mag.
NGC 5925	Open Cluster	15h 27m 42.0s	−54° 31.0′	8.4
NGC 5946	Globular Cluster	15h 35m 30.0s	−50° 40.0′	9.6
NGC 5999	Open Cluster	15h 52m 12.0s	−56° 28.0′	9.0
NGC 6031	Open Cluster	16h 07m 35.0s	−54° 00.9′	8.5
NGC 6067	Open Cluster	16h 13m 11.0s	−54° 13.1′	5.6
NGC 6087	Open Cluster	16h 18m 50.5s	−57° 56.1′	5.4
NGC 6115	Open Cluster	16h 24m 26.3s	−51° 56.9′	9.8
NGC 6134	Open Cluster	16h 27m 46.5s	−49° 09.1′	7.2
NGC 6152	Open Cluster	16h 32m 45.5s	−52° 38.6′	8.1
NGC 6167	Open Cluster	16h 34m 34.9s	−49° 46.3′	6.7
NGC 6169	Open Cluster	16h 34m 04.6s	−44° 02.7′	6.6

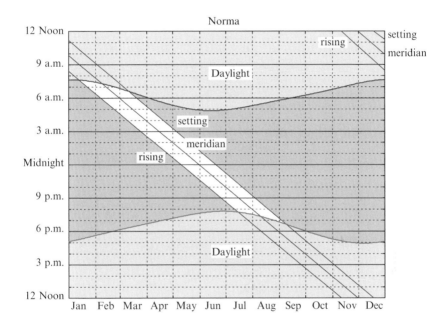

Octans

the Octant

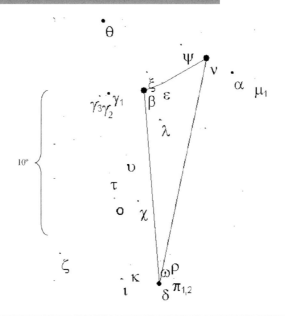

Star	Right ascension	Declination	Mag.
α	21h 04m 42.8s	−77° 01′ 26″	5.15
β	22h 46m 03.3s	−81° 22′ 54″	4.15
γ₁	00h 10m 02.3s	−82° 13′ 27″	5.28
γ₂	23h 52m 06.8s	−82° 01′ 08″	5.11
γ₃	23h 57m 32.8s	−82° 10′ 12″	5.73
δ	14h 26m 54.8s	−83° 40′ 04″	4.32
ε	22h 20m 01.4s	−80° 26′ 23″	5.10
ζ	08h 56m 41.9s	−85° 39′ 47″	5.42
η	10h 59m 14.0s	−84° 35′ 37″	6.19
θ	00h 01m 35.8s	−77° 03′ 57″	4.78
ι	12h 54m 58.9s	−85° 07′ 24″	5.46
κ	13h 40m 55.8s	−85° 47′ 09″	5.58
λ	21h 50m 54.2s	−82° 43′ 09″	5.29
μ₁	20h 42m 03.1s	−76° 10′ 50″	6.00
μ₂	20h 41m 43.6s	−75° 21′ 02″	6.55
ν	21h 41m 28.6s	−77° 23′ 24″	3.76
ξ	22h 50m 22.9s	−80° 07′ 27″	5.35
π₁	15h 01m 50.2s	−83° 13′ 39″	5.65
π₂	15h 04m 46.6s	−83° 02′ 17″	5.65
ρ	15h 43m 17.0s	−84° 27′ 55″	5.57
σ	21h 08m 44.8s	−88° 57′ 24″	5.47
τ	23h 28m 03.9s	−87° 28′ 57″	5.49
υ	22h 31m 37.6s	−85° 58′ 03″	5.77
φ	18h 23m 36.0s	−75° 02′ 39″	5.47
χ	18h 54m 44.3s	−87° 36′ 21″	5.28
ψ	22h 17m 50.0s	−77° 30′ 42″	5.51
ω	15h 11m 08.2s	−84° 47′ 15″	5.91

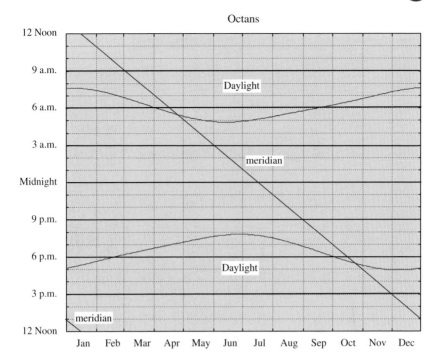

The constellation Octans is not visible from mid-northern latitudes

Ophiuchus
the Serpent Holder

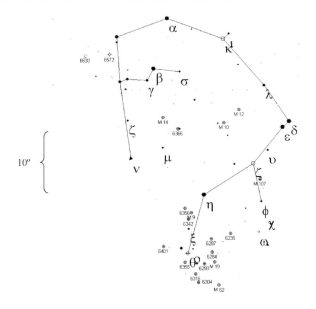

Star	Right ascension	Declination	Mag.
α	17h 34m 56.0s	+12° 33′ 36″	2.08
β	17h 43m 28.3s	+04° 34′ 02″	2.77
γ	17h 47m 53.5s	+02° 42′ 26″	3.75
δ	16h 14m 20.6s	−03° 41′ 40″	2.74
ε	16h 18m 19.2s	−04° 41′ 33″	3.24
ζ	16h 37m 09.4s	−10° 34′ 02″	2.56
η	17h 10m 22.6s	−15° 43′ 29″	2.43
θ	17h 22m 00.5s	−24° 59′ 58″	3.27
ι	16h 54m 00.4s	+10° 09′ 55″	4.38
κ	16h 57m 40.0s	+09° 22′ 30″	3.20
λ	16h 30m 54.7s	+01° 59′ 02″	3.82
μ	17h 37m 50.6s	−08° 07′ 08″	4.62
ν	17h 59m 01.5s	−09° 46′ 25″	3.34
ξ	17h 21m 00.1s	−21° 06′ 46″	4.39
o_1	17h 18m 00.6s	−24° 17′ 13″	5.20
o_2	17h 18m 00.4s	−24° 17′ 03″	6.80
$ρ_1$	16h 25m 35.1s	−23° 26′ 50″	5.02
$ρ_2$	16h 25m 35.0s	−23° 26′ 46″	5.92
σ	17h 26m 30.8s	+04° 08′ 25″	4.34
$τ_1$	18h 03m 04.8s	−08° 10′ 50″	5.94
$τ_2$	18h 03m 04.8s	−08° 10′ 50″	5.24
υ	16h 27m 48.1s	−08° 22′ 18″	4.63
φ	16h 31m 08.2s	−16° 36′ 46″	4.28
χ	16h 27m 01.3s	−18° 27′ 23″	4.42
ψ	16h 24m 06.1s	−20° 02′ 15″	4.50
ω	16h 32m 08.0s	−21° 27′ 59″	4.45

Deep sky object	Description	Right ascension	Declination	Mag.
M9	Globular Cluster	17h 19m 12.0s	−18° 31.0′	7.6
M10	Globular Cluster	16h 57m 06.0s	−04° 06.0′	6.6
M12	Globular Cluster	16h 47m 12.0s	−01° 57.0′	6.8
M14	Globular Cluster	17h 37m 36.0s	−03° 15.0′	7.6
M19	Globular Cluster	17h 02m 36.0s	−26° 16.0′	6.7
M62	Globular Cluster	17h 01m 12.0s	−30° 07.0′	6.7
M107	Globular Cluster	16h 32m 31.9s	−13° 03.2′	8.1
NGC 6235	Globular Cluster	16h 53m 24.0s	−22° 11.0′	10.0
NGC 6284	Globular Cluster	17h 04m 30.0s	−24° 46.0′	8.9
NGC 6287	Globular Cluster	17h 05m 12.0s	−22° 42.0′	9.3
NGC 6293	Globular Cluster	17h 10m 12.0s	−26° 35.0′	8.2
NGC 6304	Globular Cluster	17h 14m 30.0s	−29° 28.0′	8.4
NGC 6316	Globular Cluster	17h 16m 36.0s	−28° 08.0′	8.8
NGC 6342	Globular Cluster	17h 21m 12.0s	−19° 35.0′	9.8
NGC 6355	Globular Cluster	17h 24m 00.0s	−26° 21.0′	9.7
NGC 6356	Globular Cluster	17h 23m 36.0s	−17° 49.0′	8.2
NGC 6366	Globular Cluster	17h 27m 42.0s	−05° 05.0′	8.9
NGC 6401	Globular Cluster	17h 38m 36.0s	−23° 55.0′	9.5
NGC 6572	Planetary Nebula	18h 12m 06.4s	+06° 51.2′	8.1
NGC 6633	Open Cluster	18h 27m 42.0s	+06° 34.0′	4.6

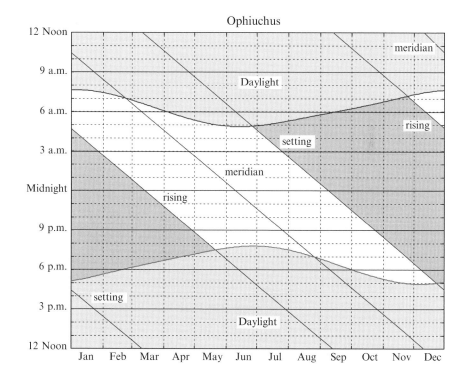

Orion

the Great Hunter

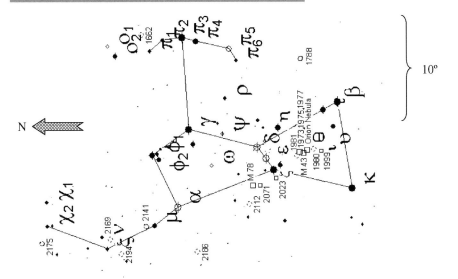

Star	Right ascension	Declination	Mag.
α	5h 55m 10.3s	+07° 24′ 25″	0.50
β	5h 14m 32.2s	−08° 12′ 06″	0.12
γ	5h 25m 07.8s	+06° 20′ 59″	1.64
δ_1	5h 32m 00.4s	−00° 17′ 04″	6.85
δ_2	5h 32m 00.3s	−00° 17′ 57″	2.23
ε	5h 36m 12.7s	−01° 12′ 07″	1.70
ζ_1	5h 40m 45.5s	−01° 56′ 34″	2.05
ζ_2	5h 40m 45.5s	−01° 56′ 34″	4.20
η	5h 24m 28.6s	−02° 23′ 49″	3.36
θ_1	5h 35m 15.8s	−05° 23′ 14″	6.73
θ_2	5h 35m 16.0s	−05° 23′ 07″	7.96
θ_3	5h 35m 16.4s	−05° 23′ 23″	5.13
θ_4	5h 35m 17.2s	−05° 23′ 16″	6.70
θ_5	5h 35m 22.8s	−05° 24′ 58″	6.39
ι	5h 35m 25.9s	−05° 54′ 36″	2.77
κ	5h 47m 45.3s	−09° 40′ 11″	2.06
λ_1	5h 35m 08.2s	+09° 56′ 03″	3.66
λ_2	5h 35m 08.4s	+09° 56′ 06″	5.56
μ	6h 02m 23.0s	+09° 38′ 51″	4.12
ν	6h 07m 34.3s	+14° 46′ 06″	4.42
ξ	6h 11m 56.4s	+14° 12′ 31″	4.48
o_1	4h 52m 31.9s	+14° 15′ 02″	4.74
o_2	4h 56m 22.2s	+13° 30′ 52″	4.07
π_1	4h 54m 53.7s	+10° 09′ 03″	4.65
π_2	4h 50m 36.7s	+08° 54′ 01″	4.36
π_3	4h 49m 50.3s	+06° 57′ 41″	3.19
π_4	4h 51m 12.3s	+05° 36′ 18″	3.69
π_5	4h 54m 15.0s	+02° 26′ 26″	3.72
π_6	4h 58m 32.8s	+01° 42′ 51″	4.47
ρ	5h 13m 17.4s	+02° 51′ 40″	4.46

Star	Right ascension	Declination	Mag.
σ	5h 38m 44.7s	−02° 36′ 00″	3.81
τ	5h 17m 36.3s	−06° 50′ 40″	3.60
υ	5h 31m 55.8s	−07° 18′ 06″	4.62
φ_1	5h 34m 49.2s	+09° 29′ 22″	4.41
φ_2	5h 36m 54.3s	+09° 17′ 26″	4.09
χ_1	5h 54m 22.9s	+20° 16′ 34″	4.41
χ_2	6h 03m 55.2s	+20° 08′ 18″	4.63
ψ_1	5h 24m 44.8s	+01° 50′ 47″	4.95
ψ_2	5h 26m 50.2s	+03° 05′ 44″	4.59
ω	5h 39m 11.1s	+04° 07′ 17″	4.57

Deep sky object	Description	Right ascension	Declination	Mag.
M42	Great Nebula	05h 35m 16.5s	−05° 23.4′	3.7
M43	deMairan's Nebula	05h 35m 31.3s	−05° 16.1′	7.0
M78	Diffuse Nebula	05h 46m 45.8s	+00° 04.7′	8.0
NGC 1662	Open Cluster	04h 48m 28.9s	+10° 55.8′	6.4
NGC 1788	Diffuse Nebula	05h 06m 53.1s	−03° 20.5′	9.0
NGC 1973	Diffuse Nebula	05h 35m 04.8s	−04° 43.9′	7.0
NGC 1975	Diffuse Nebula	05h 35m 18.3s	−04° 41.1′	7.0
NGC 1977	Open Cluster	05h 35m 15.8s	−04° 50.7′	4.2
NGC 1980	Open Cluster	05h 35m 25.9s	−05° 54.6′	2.5
NGC 1981	Open Cluster	05h 35m 09.6s	−04° 25.5′	4.2
NGC 1999	Diffuse Nebula	05h 36m 25.3s	−06° 42.9′	9.0
NGC 2023	Diffuse Nebula	05h 41m 38.3s	−02° 15.5′	9.0
NGC 2071	Diffuse Nebula	05h 47m 07.3s	+00° 17.6′	8.0
NGC 2112	Open Cluster	05h 53m 45.2s	+00° 24.6′	9.1
NGC 2141	Open Cluster	06h 02m 55.1s	+10° 26.8′	9.4
NGC 2169	Open Cluster	06h 08m 24.3s	+13° 57.9′	5.9
NGC 2175	Open Cluster	06h 09m 39.6s	+20° 29.2′	6.8
NGC 2186	Open Cluster	06h 12m 07.1s	+05° 27.5′	8.7
NGC 2194	Open Cluster	06h 13m 45.9s	+12° 48.4′	8.5

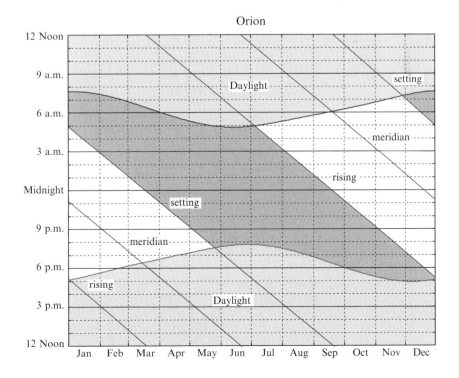

Orion

Pavo

the Peacock

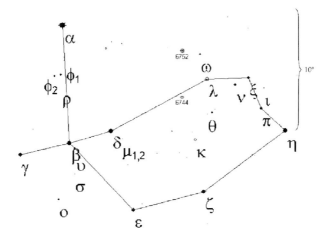

Star	Right ascension	Declination	Mag.
α	20h 25m 38.8s	−56° 44′ 07″	1.94
β	20h 44m 57.4s	−66° 12′ 11″	3.42
γ	21h 26m 26.7s	−65° 21′ 59″	4.22
δ	20h 08m 43.2s	−66° 10′ 56″	3.56
ε	20h 00m 35.4s	−72° 54′ 38″	3.96
ζ	18h 43m 02.1s	−71° 25′ 42″	4.01
η	17h 45m 43.9s	−64° 43′ 26″	3.62
θ	18h 48m 37.8s	−65° 04′ 40″	5.73
ι	18h 10m 26.1s	−62° 00′ 08″	5.49
κ	18h 56m 57.0s	−67° 14′ 01″	4.44
λ	18h 52m 12.9s	−62° 11′ 16″	4.22
μ_1	20h 00m 22.9s	−66° 56′ 58″	5.76
μ_2	20h 01m 52.3s	−66° 56′ 39″	5.31
ν	18h 31m 22.3s	−62° 16′ 42″	4.64
ξ	18h 23m 13.5s	−61° 29′ 38″	4.36
o	21h 13m 20.4s	−70° 07′ 36″	5.02
π	18h 08m 34.7s	−63° 40′ 06″	4.35
ρ	20h 37m 35.2s	−61° 31′ 48″	4.88
σ	20h 49m 18.0s	−68° 46′ 36″	5.41
τ	19h 16m 28.5s	−69° 11′ 26″	6.27
υ	20h 41m 57.1s	−66° 45′ 39″	5.15
φ_1	20h 35m 34.7s	−60° 34′ 54″	4.76
φ_2	20h 40m 02.4s	−60° 32′ 56″	5.12
ω	18h 58m 36.3s	−60° 12′ 02″	5.14

Deep sky object	Description	Right ascension	Declination	Mag.
NGC 6744	Galaxy	19h 09m 48.0s	−63° 51.0′	9.2
NGC 6752	Globular Cluster	19h 10m 54.0s	−59° 59.0′	5.4

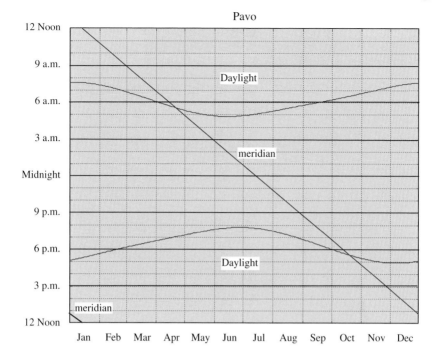
Pavo

The constellation Pavo is not visible from mid-northern latitudes

Pegasus
the Flying Horse

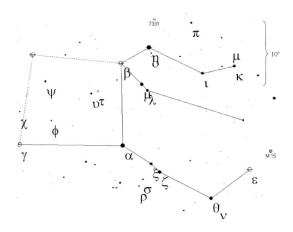

Star	Right ascension	Declination	Mag.
α	23h 04m 45.6s	+15° 12′ 19″	2.49
β	23h 03m 46.4s	+28° 04′ 58″	2.42
γ	00h 13m 14.1s	+15° 11′ 01″	2.83
ε	21h 44m 11.1s	+09° 52′ 30″	2.39
ζ	22h 41m 27.6s	+10° 49′ 53″	3.40
η	22h 43m 00.1s	+30° 13′ 17″	2.94
θ	22h 10m 11.9s	+06° 11′ 52″	3.53
ι	22h 07m 00.6s	+25° 20′ 42″	3.76
κ	21h 44m 38.6s	+25° 38′ 42″	4.13
λ	22h 46m 31.8s	+23° 33′ 56″	3.95
μ	22h 50m 00.1s	+24° 36′ 06″	3.48
ν	22h 05m 40.7s	+05° 03′ 31″	4.84
ξ	22h 46m 41.5s	+12° 10′ 22″	4.19
ο	22h 41m 45.3s	+29° 18′ 27″	4.79
π_1	22h 09m 13.5s	+33° 10′ 21″	5.58
π_2	22h 09m 59.2s	+33° 10′ 42″	4.29
ρ	22h 55m 13.6s	+08° 48′ 58″	4.90
σ	22h 52m 24.0s	+09° 50′ 09″	5.16
τ	23h 20m 38.2s	+23° 44′ 25″	4.60
υ	23h 25m 22.7s	+23° 24′ 15″	4.40
φ	23h 52m 29.2s	+19° 07′ 13″	5.08
χ	00h 14m 36.1s	+20° 12′ 24″	4.80
ψ	23h 57m 45.5s	+25° 08′ 29″	4.66

Deep sky object	Description	Right ascension	Declination	Mag.
M15	Globular Cluster	21h 29m 58.4s	+12° 10.0′	6.0
NGC 7331	Spiral Galaxy	22h 37m 06.0s	+34° 25.0′	9.5

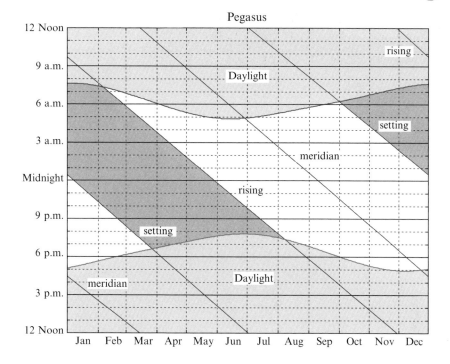

Pegasus

Perseus

the Hero

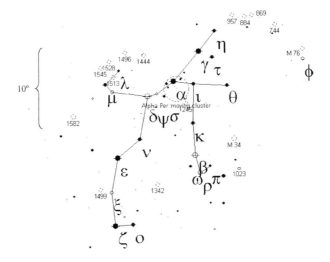

Star	Right ascension	Declination	Mag.
α	3 h 24 m 19.3 s	+49° 51′ 41″	1.79
β	3 h 08 m 10.1 s	+40° 57′ 21″	2.12
γ	3 h 04 m 47.7 s	+53° 30′ 23″	2.93
δ	3 h 42 m 55.4 s	+47° 47′ 15″	3.01
ε	3 h 57 m 51.2 s	+40° 00′ 37″	2.89
ζ	3 h 54 m 07.9 s	+31° 53′ 01″	2.85
η	2 h 50 m 41.8 s	+55° 53′ 44″	3.76
θ	2 h 44 m 11.9 s	+49° 13′ 43″	4.12
ι	3 h 09 m 04.0 s	+49° 36′ 48″	4.05
κ	3 h 09 m 29.7 s	+44° 51′ 27″	3.80
λ	4 h 06 m 35.0 s	+50° 21′ 05″	4.29
μ	4 h 14 m 53.8 s	+48° 24′ 34″	4.14
ν	3 h 45 m 11.6 s	+42° 34′ 43″	3.77
ξ	3 h 58 m 57.8 s	+35° 47′ 28″	4.04
o	3 h 44 m 19.1 s	+32° 17′ 18″	3.83
π	2 h 58 m 45.6 s	+39° 39′ 46″	4.70
ρ	3 h 05 m 10.5 s	+38° 50′ 25″	3.39
σ	3 h 30 m 34.4 s	+47° 59′ 43″	4.36
τ	2 h 54 m 15.4 s	+52° 45′ 45″	3.95
φ	1 h 43 m 39.6 s	+50° 41′ 20″	4.07
ψ	3 h 36 m 29.3 s	+48° 11′ 34″	4.23
ω	3 h 11 m 17.3 s	+39° 36′ 42″	4.63

Deep sky object	Description	Right ascension	Declination	Mag.
M34	Open Cluster	02h 42m 07.4s	+42° 44.8'	5.2
M76	Little Dumbbell Nebula	01h 42m 18.1s	+51° 34.3'	10.1
NGC 744	Open Cluster	01h 58m 29.9s	+55° 28.5'	7.9
NGC 869	part of Double Cluster	02h 19m 03.8s	+57° 08.1'	5.3
NGC 884	part of Double Cluster	02h 22m 32.1s	+57° 08.6'	6.1
NGC 957	Open Cluster	02h 33m 19.0s	+57° 34.2'	7.6
NGC 1023	Galaxy	02h 40m 23.9s	+39° 03.8'	9.2
NGC 1245	Open Cluster	03h 14m 41.5s	+47° 14.3'	8.4
NGC 1342	Open Cluster	03h 31m 40.1s	+37° 22.5'	6.7
NGC 1444	Open Cluster	03h 49m 22.9s	+52° 39.7'	6.6
NGC 1496	Open Cluster	04h 04m 31.9s	+52° 39.7'	9.6
NGC 1499	California Nebula	04h 03m 14.4s	+36° 22.1'	6.0
NGC 1513	Open Cluster	04h 09m 54.7s	+49° 31.0'	8.4
NGC 1528	Open Cluster	04h 15m 18.9s	+51° 12.7'	6.4
NGC 1545	Open Cluster	04h 20m 56.3s	+50° 15.3'	6.2
NGC 1582	Open Cluster	04h 32m 00.0s	+43° 50.9'	7.0

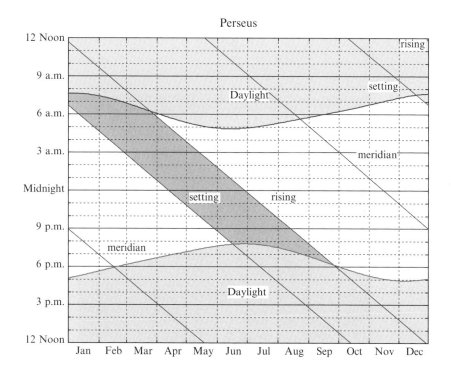

Perseus

Phoenix

the Firebird

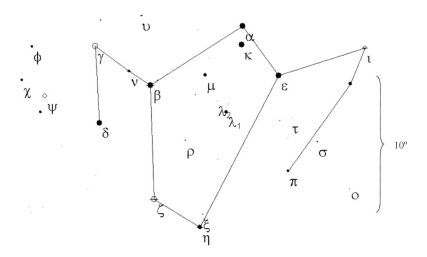

Star	Right ascension	Declination	Mag.
α	00h 26m 17.0s	−42° 18′ 22″	2.39
β	01h 06m 05.0s	−46° 43′ 08″	3.31
γ	01h 28m 21.9s	−43° 19′ 06″	3.41
δ	01h 31m 15.0s	−49° 04′ 22″	3.95
ε	00h 09m 24.6s	−45° 44′ 51″	3.88
ζ	01h 08m 23.0s	−55° 14′ 45″	3.92
η	00h 43m 21.2s	−57° 27′ 47″	4.36
θ	23h 39m 27.9s	−46° 38′ 16″	6.09
ι	23h 35m 04.5s	−42° 36′ 55″	4.71
κ	00h 26m 12.1s	−43° 40′ 48″	3.94
λ₁	00h 31m 24.9s	−48° 48′ 13″	4.77
λ₂	00h 35m 41.0s	−48° 00′ 04″	5.51
μ	00h 41m 19.5s	−46° 05′ 06″	4.59
ν	01h 15m 11.1s	−45° 31′ 54″	4.96
ξ	00h 41m 46.3s	−56° 30′ 06″	5.70
π	23h 58m 55.7s	−52° 44′ 45″	5.13
ρ	00h 50m 41.0s	−50° 59′ 13″	5.22
σ	23h 47m 15.9s	−50° 13′ 35″	5.18
τ	00h 01m 04.4s	−48° 48′ 36″	5.71
υ	01h 07m 47.8s	−41° 29′ 14″	5.21
φ	01h 54m 21.9s	−42° 29′ 50″	5.11
χ	02h 01m 42.3s	−44° 42′ 48″	5.14
ψ	01h 53m 38.7s	−46° 18′ 10″	4.41
ω	01h 02m 01.8s	−57° 00′ 09″	6.11

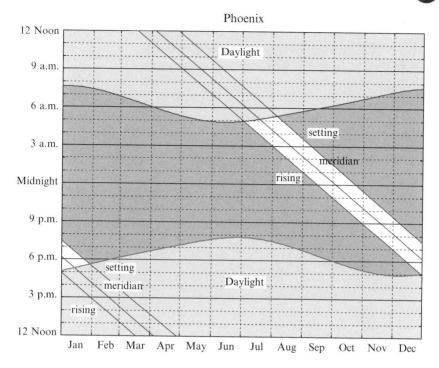
Phoenix

Pictor

the Painter's Easel

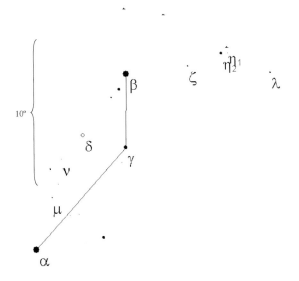

Star	Right ascension	Declination	Mag.
α	06h 48m 11.4s	−61° 56′ 29″	3.27
β	05h 47m 17.1s	−51° 03′ 59″	3.85
γ	05h 49m 49.6s	−56° 10′ 00″	4.51
δ	06h 10m 17.9s	−54° 58′ 07″	4.81
ζ	05h 19m 22.0s	−50° 36′ 22″	5.45
η₁	05h 02m 48.6s	−49° 09′ 05″	5.38
η₂	05h 04m 57.9s	−49° 34′ 41″	5.03
θ	05h 24m 46.1s	−52° 18′ 59″	6.27
ι₁	04h 50m 55.1s	−53° 27′ 41″	5.61
ι₂	04h 50m 56.3s	−53° 27′ 35″	6.42
κ	05h 22m 22.2s	−56° 08′ 04″	6.11
λ	04h 42m 46.3s	−50° 28′ 52″	5.31
μ	06h 31m 58.4s	−58° 45′ 15″	5.70
ν	06h 22m 55.8s	−56° 22′ 12″	5.61

Pictor

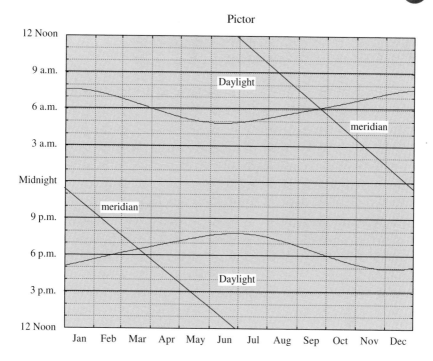

The constellation Pictor is not visible from mid-northern latitudes

Pisces

the Fishes

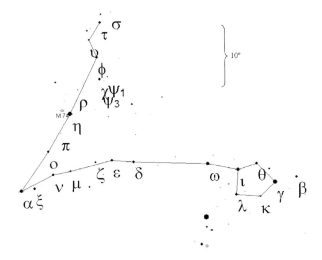

Star	Right ascension	Declination	Mag.
α_1	02h 02m 02.7s	+02° 45′ 49″	5.28
α_2	02h 02m 02.7s	+02° 45′ 49″	4.16
β	23h 03m 52.5s	+03° 49′ 12″	4.48
γ	23h 17m 09.9s	+03° 16′ 56″	3.70
δ	00h 48m 40.9s	+07° 35′ 06″	4.44
ε	01h 02m 56.5s	+07° 53′ 24″	4.27
ζ_1	01h 13m 43.8s	+07° 34′ 31″	5.28
ζ_2	01h 13m 45.2s	+07° 34′ 42″	6.43
η	01h 31m 28.9s	+15° 20′ 45″	3.62
θ	23h 27m 58.0s	+06° 22′ 44″	4.27
ι	23h 39m 57.0s	+05° 37′ 35″	4.13
κ	23h 26m 55.9s	+01° 15′ 20″	4.95
λ	23h 42m 02.7s	+01° 46′ 48″	4.49
μ	01h 30m 11.1s	+06° 08′ 38″	4.84
ν	01h 41m 25.8s	+05° 29′ 15″	4.45
ξ	01h 53m 33.3s	+03° 11′ 15″	4.61
o	01h 45m 23.6s	+09° 09′ 28″	4.26
π	01h 37m 05.9s	+12° 08′ 30″	5.54
ρ	01h 26m 15.2s	+19° 10′ 20″	5.35
σ	01h 02m 49.0s	+31° 48′ 16″	5.50
τ	01h 11m 39.6s	+30° 05′ 23″	4.51
υ	01h 19m 27.9s	+27° 15′ 51″	4.74
φ	01h 13m 44.8s	+24° 35′ 01″	4.67
χ	01h 11m 27.1s	+21° 02′ 05″	4.66
ψ_1	01h 05m 40.9s	+21° 28′ 24″	5.33
ψ_2	01h 05m 41.6s	+21° 27′ 56″	5.52
ψ_3	01h 07m 57.1s	+20° 44′ 21″	5.56
ψ_4	01h 09m 49.1s	+19° 39′ 31″	5.57
ω	23h 59m 18.6s	+06° 51′ 48″	4.03

Deep sky object	Description	Right ascension	Declination	Mag.
M74	Galaxy	01h 36m 41.6s	+15° 47.1'	9.4

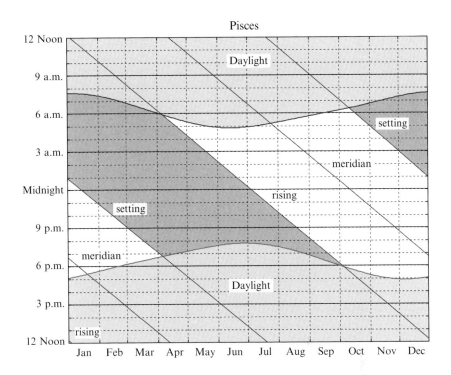

Piscis Austrinus
the Southern Fish

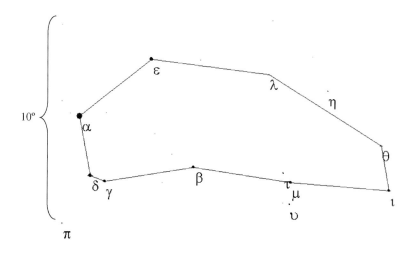

Star	Right ascension	Declination	Mag.
α	22h 57m 39.0s	−29° 37′ 20″	1.16
β	22h 31m 30.3s	−32° 20′ 46″	4.29
γ	22h 52m 31.5s	−32° 52′ 32″	4.46
δ	22h 55m 56.8s	−32° 32′ 23″	4.21
ε	22h 40m 39.3s	−27° 02′ 37″	4.17
ζ	22h 30m 53.6s	−26° 04′ 25″	6.43
η	22h 00m 50.1s	−28° 27′ 13″	5.42
θ	21h 47m 44.1s	−30° 53′ 54″	5.01
ι	21h 44m 56.7s	−33° 01′ 33″	4.34
λ	22h 14m 18.7s	−27° 46′ 01″	5.43
μ	22h 08m 22.9s	−32° 59′ 19″	4.50
π	23h 03m 29.7s	−34° 44′ 58″	5.11
τ	22h 10m 08.7s	−32° 32′ 55″	4.92
υ	22h 08m 25.9s	−34° 02′ 38″	4.99

Piscis Austrinus

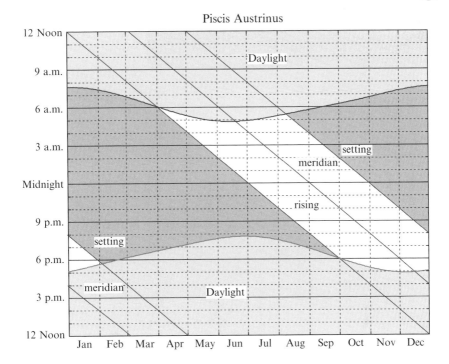

Puppis
the Ship's Stern

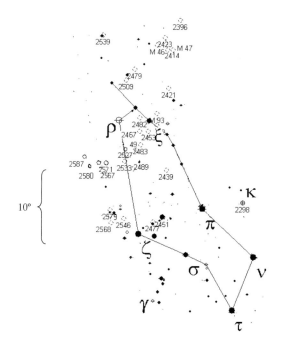

10°

Star	Right ascension	Declination	Mag.
ζ	08h 03m 35.0s	−40° 00′ 11″	2.25
ν	06h 37m 45.6s	−43° 11′ 45″	3.17
ξ	07h 49m 17.6s	−24° 51′ 35″	3.34
o₁	07h 33m 58.4s	−39° 54′ 21″	6.76
o₂	07h 48m 05.1s	−25° 56′ 14″	4.50
π	07h 17m 08.5s	−37° 05′ 51″	2.70
ρ	08h 07m 32.6s	−24° 18′ 15″	2.81
σ	07h 29m 13.8s	−43° 18′ 05″	3.25
τ	06h 49m 56.1s	−50° 36′ 53″	2.93

Deep sky object	Description	Right ascension	Declination	Mag.
M46	Open Cluster	07h 41m 46.8s	–14° 48.6′	6.1
M47	Open Cluster	07h 36m 35.0s	–14° 28.9′	4.4
M93	Open Cluster	07h 44m 29.2s	–23° 51.2′	6.2
NGC 2298	Globular Cluster	06h 48m 59.1s	–36° 00.3′	9.2
NGC 2396	Open Cluster	07h 28m 02.9s	–11° 43.2′	7.4
NGC 2414	Open Cluster	07h 33m 12.8s	–15° 27.2′	7.9
NGC 2421	Open Cluster	07h 36m 11.8s	–20° 36.7′	8.3
NGC 2423	Open Cluster	07h 37m 06.7s	–13° 52.8′	6.7
NGC 2439	Open Cluster	07h 40m 45.4s	–31° 41.5′	6.9
NGC 2451	Open Cluster	07h 45m 15.0s	–37° 58.1′	2.8
NGC 2453	Open Cluster	07h 47m 34.1s	–27° 11.7′	8.3
NGC 2467	Open Cluster	07h 52m 29.5s	–26° 25.8′	7.1
NGC 2477	Open Cluster	07h 52m 09.8s	–38° 32.0′	5.8
NGC 2479	Open Cluster	07h 55m 06.1s	–17° 42.5′	9.6
NGC 2482	Open Cluster	07h 55m 10.3s	–24° 15.3′	7.3
NGC 2483	Open Cluster	07h 55m 38.8s	–27° 53.2′	7.6
NGC 2489	Open Cluster	07h 56m 15.9s	–30° 03.8′	7.9
NGC 2509	Open Cluster	08h 00m 47.8s	–19° 03.0′	9.3
NGC 2527	Open Cluster	08h 04m 58.2s	–28° 08.8	6.5
NGC 2533	Open Cluster	08h 07m 04.1s	–29° 53.0′	7.6
NGC 2539	Open Cluster	08h 10m 37.9s	–12° 49.1′	6.5
NGC 2546	Open Cluster	08h 12m 15.6s	–37° 35.6′	6.3
NGC 2567	Open Cluster	08h 18m 29.1s	–30° 38.7′	7.4
NGC 2568	Open Cluster	08h 18m 18.1s	–37° 06.3′	10.0
NGC 2571	Open Cluster	08h 18m 56.3s	–29° 44.9′	7.0
NGC 2579	Open Cluster	08h 20m 53.0s	–36° 13.0′	7.5
NGC 2580	Open Cluster	08h 21m 27.9s	–30° 17.6′	9.7
NGC 2587	Open Cluster	08h 23m 24.1s	–29° 30.5′	9.2

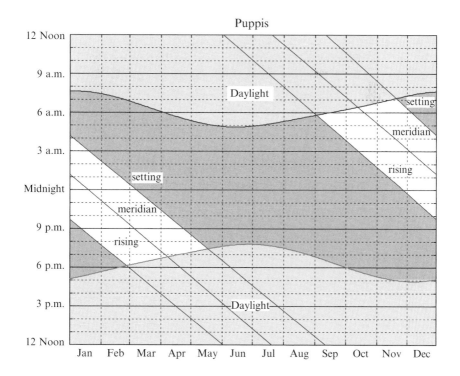

Puppis

Pyxis

the Ship's Compass

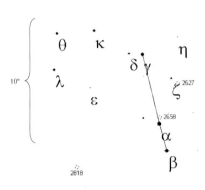

Star	Right ascension	Declination	Mag.
α	08h 43m 35.5s	−33° 11′ 11″	3.68
β	08h 40m 06.2s	−35° 18′ 29″	3.97
γ	08h 50m 31.9s	−27° 42′ 36″	4.01
δ	08h 55m 31.5s	−27° 40′ 55″	4.89
ε	09h 09m 56.4s	−30° 21′ 55″	5.59
ζ	08h 39m 42.5s	−29° 33′ 40″	4.89
η	08h 37m 52.1s	−26° 15′ 18″	5.27
θ	09h 21m 29.6s	−25° 57′ 56″	4.72
κ	09h 08m 02.8s	−25° 51′ 30″	4.58
λ	09h 23m 12.1s	−28° 50′ 02″	4.69

Deep sky object	Description	Right ascension	Declination	Mag.
NGC 2627	Open Cluster	08h 37m 14.9s	−29° 57.0′	8.4
NGC 2658	Open Cluster	08h 43m 27.3s	−32° 39.4′	9.2
NGC 2818	Open Cluster	09h 16m 00.0s	−36° 37.0′	8.2

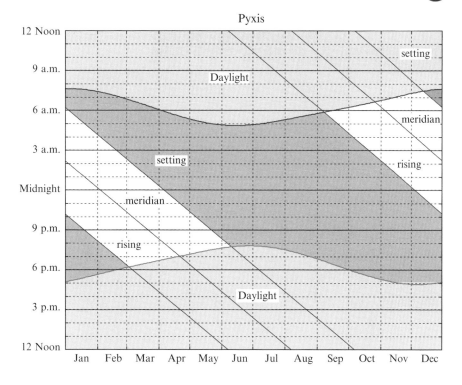

Pyxis

Reticulum

the Reticle

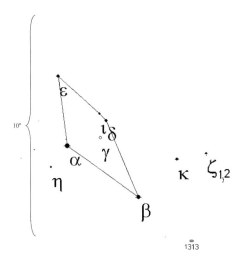

10°

1313

Star	Right ascension	Declination	Mag.
α	04h 14m 25.5s	−62° 28′ 26″	3.35
β	03h 44m 12.0s	−64° 48′ 26″	3.85
γ	04h 00m 53.8s	−62° 09′ 34″	4.51
δ	03h 58m 44.7s	−61° 24′ 01″	4.56
ε	04h 16m 28.9s	−59° 18′ 07″	4.44
ζ_1	03h 17m 46.1s	−62° 34′ 32″	5.54
ζ_2	03h 18m 12.8s	−62° 30′ 23″	5.24
η	04h 21m 53.4s	−63° 23′ 11″	5.24
θ	04h 17m 39.7s	−63° 15′ 19″	5.87
ι	04h 01m 18.2s	−61° 04′ 44″	4.97
κ	03h 29m 22.6s	−62° 56′ 16″	4.72

Deep sky object	Description	Right ascension	Declination	Mag.
NGC 1313	Galaxy	03h 18m 16.0s	−66° 29.9′	9.8

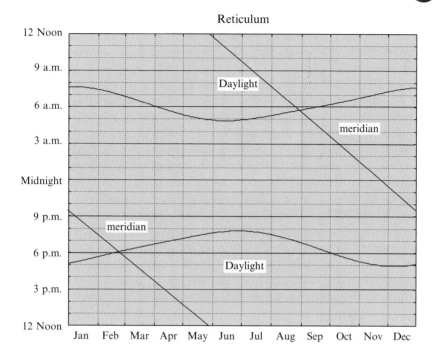

The constellation Reticulum is not visible from mid-northern latitudes

Sagitta
the Arrow

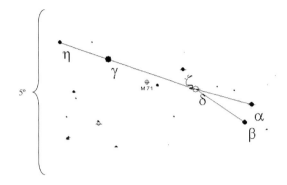

Star	Right ascension	Declination	Mag.
α	19h 40m 05.7s	+18° 00′ 50″	4.37
β	19h 41m 02.9s	+17° 28′ 33″	4.37
γ	19h 58m 45.3s	+19° 29′ 32″	3.47
δ	19h 47m 23.2s	+18° 32′ 03″	3.82
ε	19h 37m 17.3s	+16° 27′ 46″	5.66
ζ	19h 48m 58.6s	+19° 08′ 31″	5.00
η	20h 05m 09.4s	+19° 59′ 28″	5.10
θ	20h 09m 52.2s	+20° 53′ 48″	6.48

Deep sky object	Description	Right ascension	Declination	Mag.
M71	Globular Cluster	19h 53m 48.0s	+18° 47.0′	8.0

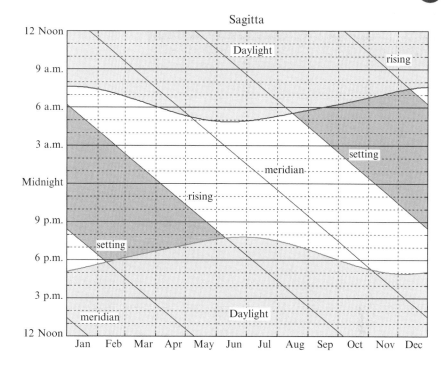

Sagitta

Sagittarius

the Archer

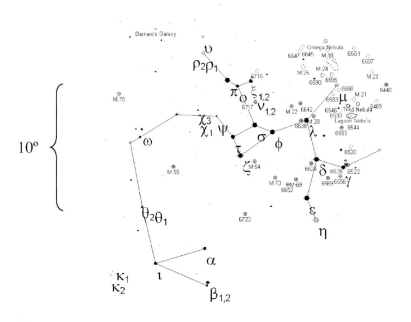

Star	Right ascension	Declination	Mag.	Star	Right ascension	Declination	Mag.
α	19h 23m 53.0s	−40° 36′ 58″	3.97	v_1	18h 54m 10.1s	−22° 44′ 42″	4.83
$β_1$	19h 22m 38.2s	−44° 27′ 32″	4.01	v_2	18h 55m 07.0s	−22° 40′ 17″	4.99
$β_2$	19h 23m 13.1s	−44° 47′ 59″	4.29	$ξ_1$	18h 57m 20.4s	−20° 39′ 23″	5.08
$γ_1$	18h 05m 01.2s	−29° 34′ 48″	4.69	$ξ_2$	18h 57m 43.7s	−21° 06′ 24″	3.51
$γ_2$	18h 05m 48.4s	−30° 25′ 27″	2.99	o	19h 04m 40.9s	−21° 44′ 30″	3.77
δ	18h 20m 59.6s	−29° 49′ 41″	2.70	π	19h 09m 45.7s	−21° 01′ 25″	2.89
ε	18h 24m 10.3s	−34° 23′ 05″	1.85	$ρ_1$	19h 21m 40.3s	−17° 50′ 50″	3.93
ζ	19h 02m 36.6s	−29° 52′ 49″	2.60	$ρ_2$	19h 21m 50.8s	−18° 18′ 30″	5.87
η	18h 17m 37.5s	−36° 45′ 42″	3.11	σ	18h 55m 15.8s	−26° 17′ 48″	2.02
$θ_1$	19h 59m 44.1s	−35° 16′ 35″	4.37	τ	19h 06m 56.3s	−27° 40′ 14″	3.32
$θ_2$	19h 59m 51.2s	−34° 41′ 53″	5.30	υ	19h 21m 43.5s	−15° 57′ 18″	4.61
ι	19h 55m 15.5s	−41° 52′ 06″	4.13	φ	18h 45m 39.3s	−26° 59′ 27″	3.17
$κ_1$	20h 22m 27.4s	−42° 02′ 59″	5.59	$χ_1$	19h 25m 16.4s	−24° 30′ 31″	5.03
$κ_2$	20h 23m 53.1s	−42° 25′ 23″	5.64	$χ_2$	19h 25m 29.6s	−23° 57′ 44″	5.43
λ	18h 27m 58.1s	−25° 25′ 18″	2.81	ψ	19h 15m 32.3s	−25° 15′ 24″	4.85
μ	18h 13m 45.7s	−21° 03′ 32″	3.86	ω	19h 55m 50.3s	−26° 17′ 58″	4.70

Deep sky object	Description	Right ascension	Declination	Mag.	Deep sky object	Description	Right ascension	Declination	Mag.
M8	Lagoon Nebula	18h 03m 48.0s	−24° 23.0′	5.0	NGC 6530	Open Cluster	18h 04m 48.0s	−24° 20.0′	4.6
M17	Omega Nebula	18h 20m 48.0s	−16° 11.0′	6.0	NGC 6544	Globular Cluster	18h 07m 18.0s	−25° 00.0′	8.1
M18	Open Cluster	18h 19m 54.0s	−17° 08.0′	6.9	NGC 6546	Open Cluster	18h 07m 12.0s	−23° 20.0′	8.0
M20	Trifid Nebula	18h 02m 36.0s	−23° 02.0′	9.0	NGC 6553	Globular Cluster	18h 09m 18.0s	−25° 54.0′	8.1

Deep sky object	Description	Right ascension	Declination	Mag.
M21	Open Cluster	18h 04m 36.0s	–22° 30.0′	5.9
M22	Globular Cluster	18h 36m 24.0s	–23° 54.0′	5.1
M23	Open Cluster	17h 56m 48.0s	–19° 01.0′	5.5
M24	Milky Way Patch	18h 18m 24.0s	–18° 25.0′	4.6
M25	Open Cluster	18h 31m 36.0s	–19° 15.0′	6.5
M28	Globular Cluster	18h 24m 30.0s	–24° 52.0′	6.8
M54	Globular Cluster	18h 55m 06.0s	–30° 29.0′	7.6
M55	Globular Cluster	19h 40m 00.0s	–30° 58.0′	6.4
M69	Globular Cluster	18h 31m 24.0s	–32° 21.0′	7.6
M70	Globular Cluster	18h 43m 12.0s	–32° 18.0′	8.0
M75	Globular Cluster	20h 06m 06.0s	–21° 55.0′	8.5
NGC 6440	Globular Cluster	17h 48m 54.0s	–20° 22.0′	9.1
NGC 6469	Open Cluster	17h 52m 54.0s	–22° 21.0′	8.2
NGC 6507	Open Cluster	17h 59m 36.0s	–17° 24.0′	9.6
NGC 6520	Open Cluster	18h 03m 24.0s	–27° 54.0′	7.6
NGC 6522	Globular Cluster	18h 03m 35.0s	–30° 02.1′	8.4
NGC 6528	Globular Cluster	18h 04m 48.0s	–30° 03.0′	9.5

Deep sky object	Description	Right ascension	Declination	Mag.
NGC 6558	Globular Cluster	18h 10m 18.0s	–31° 46.0′	9.8
NGC 6561	Open Cluster	18h 10m 30.0s	–16° 48.0′	9.0
NGC 6568	Open Cluster	18h 12m 48.0s	–21° 36.0′	8.6
NGC 6569	Globular Cluster	18h 13m 36.0s	–31° 50.0′	8.7
NGC 6583	Open Cluster	18h 15m 48.0s	–22° 08.0′	10.0
NGC 6590	Open Cluster	18h 17m 00.0s	–19° 53.0′	7.0
NGC 6624	Globular Cluster	18h 23m 42.0s	–30° 22.0′	8.0
NGC 6638	Globular Cluster	18h 30m 54.0s	–25° 30.0′	9.1
NGC 6642	Globular Cluster	18h 31m 54.0s	–23° 29.0′	9.4
NGC 6645	Open Cluster	18h 32m 36.0s	–16° 54.0′	8.5
NGC 6647	Open Cluster	18h 31m 30.0s	–17° 21.0′	8.0
NGC 6652	Globular Cluster	18h 35m 48.0s	–32° 59.0′	8.8
NGC 6716	Open Cluster	18h 54m 36.0s	–19° 53.0′	7.5
NGC 6717	Globular Cluster	18h 55m 06.0s	–22° 42.0′	9.2
NGC 6723	Globular Cluster	18h 59m 36.0s	–36° 38.0′	7.2
NGC 6822	Barnard's Galaxy	19h 44m 54.0s	–14° 48.0′	9.3

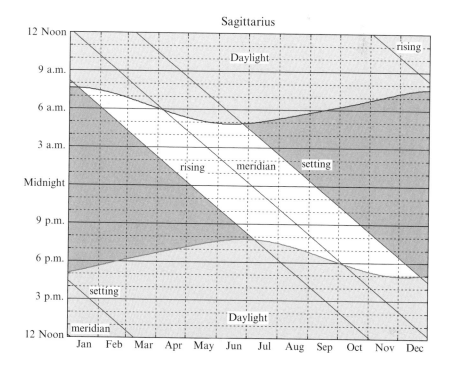

Sagittarius

Scorpius
the Scorpion

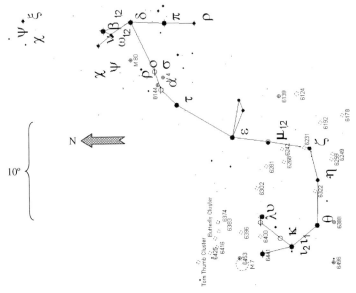

Star	Right ascension	Declination	Mag.
α	16h 29m 24.4s	−26° 25′ 55″	0.96
β₁	16h 05m 26.1s	−19° 48′ 19″	2.62
β₂	16h 05m 26.4s	−19° 48′ 07″	4.92
δ	16h 00m 19.9s	−22° 37′ 18″	2.32
ε	16h 50m 09.7s	−34° 17′ 36″	2.29
ζ₁	16h 54m 34.9s	−42° 21′ 41″	3.62
ζ₂	16h 53m 59.6s	−42° 21′ 43″	4.73
η	17h 12m 09.1s	−43° 14′ 21″	3.33
θ	17h 37m 19.0s	−42° 59′ 52″	1.87
ι₁	17h 47m 35.0s	−40° 07′ 37″	3.03
ι₂	17h 50m 11.0s	−40° 05′ 26″	4.81
κ	17h 42m 29.1s	−39° 01′ 48″	2.41
λ	17h 33m 36.4s	−37° 06′ 13″	1.63
μ₁	16h 51m 52.1s	−38° 02′ 51″	3.08
μ₂	16h 52m 20.0s	−38° 01′ 03″	3.57
ν₁	16h 11m 59.6s	−19° 27′ 38″	4.01
ν₂	16h 11m 58.5s	−19° 26′ 59″	6.30
ξ₁	16h 04m 22.0s	−11° 22′ 23″	4.77
ξ₂	16h 04m 22.0s	−11° 22′ 23″	5.07
o	16h 20m 38.1s	−24° 10′ 10″	4.55
π	15h 58m 51.0s	−26° 06′ 51″	2.89
ρ	15h 56m 53.0s	−29° 12′ 50″	3.88
σ	16h 21m 11.2s	−25° 35′ 34″	2.89
τ	16h 35m 52.9s	−28° 12′ 58″	2.82
υ	17h 30m 45.7s	−37° 17′ 45″	2.69
χ	16h 13m 50.7s	−11° 50′ 16″	5.22
ψ	16h 11m 59.9s	−10° 03′ 51″	4.94
ω₁	16h 06m 48.3s	−20° 40′ 09″	3.96
ω₂	16h 07m 24.2s	−20° 52′ 07″	4.32

Deep sky object	Description	Right ascension	Declination	Mag.
M4	Globular Cluster	16h 23m 35.4s	26° 31.5′	5.8
M6	Butterfly Cluster	17h 40m 06.0s	−32° 13.0′	4.2
M7	Ptolemy's Cluster	17h 53m 54.0s	−34° 49.0′	3.3
M80	Globular Cluster	16h 17m 03.1s	−22° 58.5′	7.3
NGC 6124	Open Cluster	16h 25m 20.0s	−40° 39.2′	5.8
NGC 6139	Globular Cluster	16h 27m 40.4s	−38° 50.9′	8.9
NGC 6144	Globular Cluster	16h 27m 14.1s	−26° 01.5′	9.0
NGC 6178	Open Cluster	16h 35m 47.2s	−45° 38.6′	7.2
NGC 6192	Open Cluster	16h 40m 23.8s	−43° 22.0′	8.5
NGC 6231	Open Cluster	16h 54m 0.0s	−41° 48.0′	2.6
NGC 6242	Open Cluster	16h 55m 36.0s	−39° 30.0′	6.4
NGC 6249	Open Cluster	16h 57m 36.0s	−44° 47.0′	8.2
NGC 6259	Open Cluster	17h 00m 42.0s	−44° 40.0′	8.0
NGC 6268	Open Cluster	17h 02m 24.0s	−39° 44.0′	9.5
NGC 6281	Open Cluster	17h 04m 48.0s	−37° 54.0′	5.4
NGC 6302	Planetary Nebula	17h 13m 44.1s	−37° 06.2′	9.6
NGC 6322	Open Cluster	17h 18m 30.0s	−42° 57.0′	6.0
NGC 6374	Open Cluster	17h 32m 18.0s	−32° 36.0′	9.0
NGC 6383	Open Cluster	17h 34m 48.0s	−32° 34.0′	5.5
NGC 6388	Globular Cluster	17h 36m 18.0s	−44° 44.0′	6.7
NGC 6396	Open Cluster	17h 38m 06.0s	−35° 00.0′	8.5
NGC 6400	Open Cluster	17h 40m 48.0s	−36° 57.0′	8.8
NGC 6416	Open Cluster	17h 44m 24.0s	−32° 21.0′	5.7
NGC 6425	Open Cluster	17h 46m 54.0s	−31° 32.0′	7.2
NGC 6441	Globular Cluster	17h 50m 12.0s	−37° 03.0′	7.2
NGC 6451	Tom Thumb Cluster	17h 50m 42.0s	−30° 13.0′	8.2
NGC 6453	Globular Cluster	17h 50m 54.0s	−34° 36.0′	9.8
NGC 6496	Globular Cluster	17h 59m 00.0s	−44° 16.0′	8.5

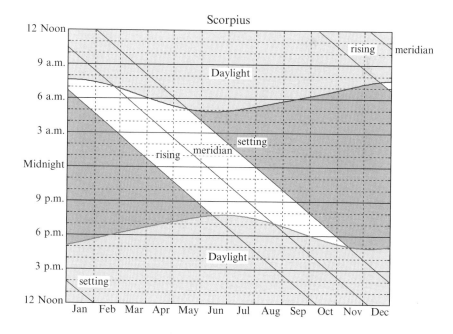

Sculptor
the Sculptor's Workshop

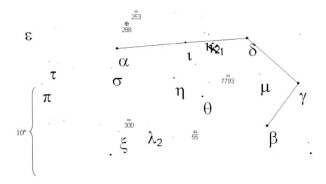

Star	Right ascension	Declination	Mag.
α	0h 58m 36.3s	−29° 21′ 28″	4.31
β	23h 32m 58.2s	−37° 49′ 07″	4.37
γ	23h 18m 49.4s	−32° 31′ 55″	4.41
δ	23h 48m 55.5s	−28° 07′ 49″	4.57
ε	01h 45m 38.7s	−25° 03′ 10″	5.31
ζ	00h 02m 19.9s	−29° 43′ 15″	5.01
η	00h 27m 55.7s	−33° 00′ 26″	4.81
θ	00h 11m 43.9s	−35° 07′ 59″	5.25
ι	00h 21m 31.2s	−28° 58′ 54″	5.18
κ_1	00h 09m 20.9s	−27° 59′ 16″	5.42
κ_2	00h 11m 34.4s	−27° 47′ 59″	5.41
λ_1	00h 42m 42.8s	−38° 27′ 48″	6.06
λ_2	00h 44m 12.0s	−38° 25′ 18″	5.90
μ	23h 40m 38.1s	−32° 04′ 24″	5.31
ξ	01h 01m 18.3s	−38° 54′ 59″	5.59
π	01h 42m 08.5s	−32° 19′ 37″	5.25
σ	01h 02m 26.3s	−31° 33′ 07″	5.50
τ	01h 36m 08.3s	−29° 54′ 27″	5.69

Deep sky object	Description	Right ascension	Declination	Mag.
NGC 55	Spiral Galaxy	00h 15m 08.4s	−39° 13.2′	7.9
NGC 253	Silver Coin Galaxy	00h 47m 33.1s	−25° 17.3′	7.1
NGC 288	Globular Cluster	00h 52m 47.4s	−26° 35.4′	8.1
NGC 300	Galaxy	00h 54m 53.4s	−37° 41.0′	8.1
NGC 7793	Galaxy	23h 57m 48.0s	−32° 35.0′	9.1

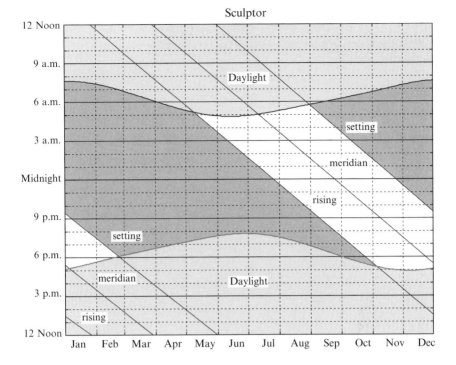

Sculptor

Scutum

the Shield

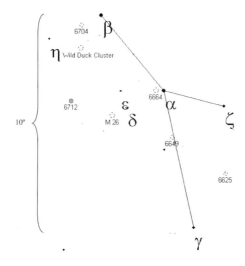

Star	Right ascension	Declination	Mag.
α	18h 35m 12.3s	−08° 14′ 39″	3.85
β	18h 47m 10.4s	−04° 44′ 53″	4.22
γ	18h 29m 11.7s	−14° 33′ 57″	4.70
δ	18h 42m 16.3s	−09° 03′ 09″	4.72
ε	18h 43m 31.2s	−08° 16′ 31″	4.90
ζ	18h 23m 39.4s	−08° 56′ 04″	4.68
η	18h 57m 03.6s	−05° 50′ 46″	4.83

Deep sky object	Description	Right ascension	Declination	Mag.
M11	Wild Duck Cluster	18h 51m 06.0s	−06° 16.0′	5.8
M26	Open Cluster	18h 45m 12.0s	−09° 24.0′	8.0
NGC 6625	Open Cluster	18h 23m 12.0s	−12° 03.0′	9.0
NGC 6649	Open Cluster	18h 33m 30.0s	−10° 24′	8.9
NGC 6664	Open Cluster	18h 36m 42.0s	−08° 13.0′	7.8
NGC 6704	Open Cluster	18h 50m 54.0s	−05° 12.0′	9.2
NGC 6712	Globular Cluster	18h 53m 06.0s	−08° 42.0′	8.2

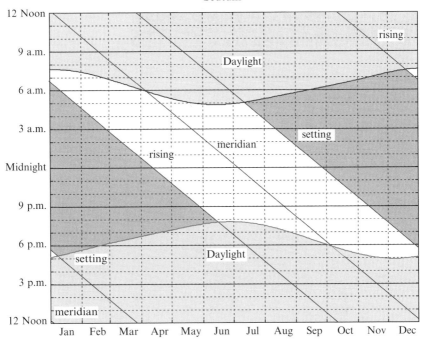

Serpens Caput
the Serpent's Head

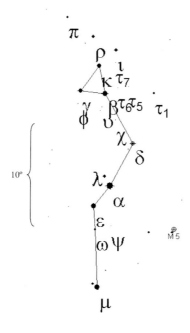

Star	Right ascension	Declination	Mag.
α	15h 44m 16.0s	+06° 25′ 32″	2.65
β	15h 46m 11.2s	+15° 25′ 18″	3.67
γ	15h 56m 27.1s	+15° 39′ 42″	3.85
δ₁	15h 34m 48.0s	+10° 32′ 15″	3.80
δ₂	15h 34m 48.1s	+10° 32′ 21″	3.80
ε	15h 50m 48.9s	+04° 28′ 40″	3.71
ι	15h 41m 33.0s	+19° 40′ 13″	4.52
κ	15h 48m 44.3s	+18° 08′ 29″	4.09
λ	15h 46m 26.5s	+07° 21′ 11″	4.43
μ	15h 49m 37.1s	−03° 25′ 49″	3.53
π	16h 02m 17.7s	+22° 48′ 16″	4.83
ρ	15h 51m 15.8s	+20° 58′ 40″	4.76
σ	16h 22m 04.3s	+01° 01′ 45″	4.82
τ₁	15h 25m 47.3s	+15° 25′ 41″	5.17
τ₂	15h 32m 09.6s	+16° 03′ 22″	6.22
τ₃	15h 35m 33.1s	+17° 39′ 20″	6.12
τ₄	15h 36m 29.2s	+16° 07′ 08″	5.93
τ₅	15h 40m 59.1s	+16° 01′ 29″	6.01
τ₆	15h 41m 54.6s	+18° 27′ 50″	5.81
τ₇	15h 44m 42.1s	+17° 15′ 51″	6.14
υ	15h 47m 17.2s	+14° 06′ 55″	5.71
φ	15h 57m 14.5s	+14° 24′ 52″	5.54
χ	15h 41m 47.3s	+12° 50′ 50″	5.33
ψ	15h 44m 01.6s	+02° 30′ 54″	5.88
ω	15h 50m 17.5s	+02° 11′ 47″	5.23

Deep sky object	Description	Right ascension	Declination	Mag.
M5	Globular Cluster	15h 18m 36.0s	+02° 05.0′	5.7

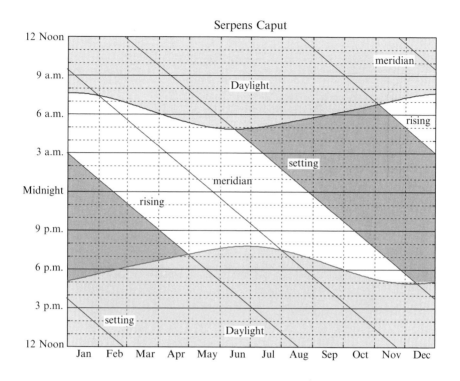

Serpens Cauda
the Serpent's Tail

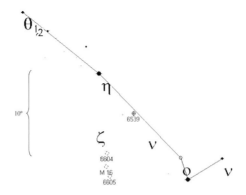

Star	Right ascension	Declination	Mag.
ζ	18h 00m 28.8s	−03° 41′ 25″	4.62
η	18h 21m 18.5s	−02° 53′ 56″	3.26
θ₁	18h 56m 13.1s	+04° 12′ 13″	4.62
θ₂	18h 56m 14.5s	+04° 12′ 07″	4.98
ν	17h 20m 49.5s	−12° 50′ 48″	4.33
ξ	17h 37m 35.1s	−15° 23′ 55″	3.54
o	17h 41m 24.8s	−12° 52′ 31″	4.26

Deep sky object	Description	Right ascension	Declination	Mag.
M16	Open Cluster	18h 18m 48.0s	−13° 47.0′	6.0
NGC 6539	Globular Cluster	18h 04m 48.0s	−07° 35.0′	9.8
NGC 6604	Open Cluster	18h 18m 06.0s	−12° 14.0′	6.5
NGC 6605	Open Cluster	18h 17m 06.0s	−14° 58.0′	6.0

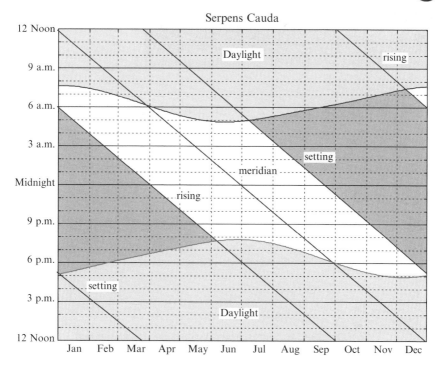

Serpens Cauda

Sextans

the Sextant

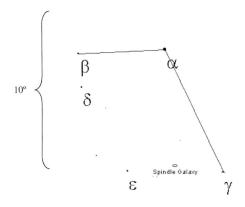

Star	Right ascension	Declination	Mag.
α	10h 07m 56.2s	–00° 22′ 18″	4.49
β	10h 30m 17.4s	–00° 38′ 13″	5.09
γ	9h 52m 30.4s	–08° 06′ 18″	5.05
δ	10h 29m 28.6s	–02° 44′ 21″	5.21
ε	10h 17m 37.7s	–08° 04′ 08″	5.24

Deep sky object	Description	Right ascension	Declination	Mag.
NGC 3115	The Spindle Galaxy	10h 05m 14.1s	–07° 43.1′	9.1

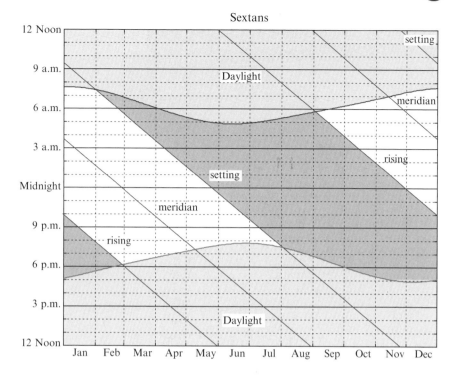

Sextans

Taurus

the Bull

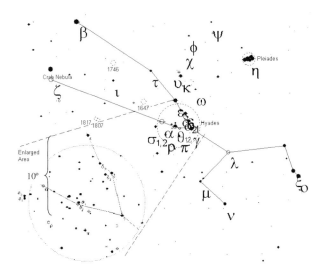

Star	Right ascension	Declination	Mag.
α	4h 35m 55.2s	+16° 30′ 33″	0.85
β	5h 26m 17.5s	+28° 36′ 27″	1.65
γ	4h 19m 47.5s	+15° 37′ 39″	3.65
δ_1	4h 22m 56.0s	+17° 32′ 33″	3.76
δ_2	4h 24m 05.7s	+17° 26′ 38″	4.80
δ_3	4h 25m 29.3s	+17° 55′ 41″	4.29
ε	4h 28m 36.9s	+19° 10′ 49″	3.53
ζ	5h 37m 38.6s	+21° 08′ 33″	3.00
η	3h 47m 29.0s	+24° 06′ 18″	2.87
θ_1	4h 28m 34.4s	+15° 57′ 44″	3.84
θ_2	4h 28m 39.7s	+15° 52′ 15″	3.40
ι	5h 03m 05.7s	+21° 35′ 24″	4.64
κ_1	4h 25m 22.1s	+22° 17′ 38″	4.22
κ_2	4h 25m 24.9s	+22° 11′ 59″	5.28
λ	4h 00m 40.8s	+12° 29′ 25″	3.47
μ	4h 15m 32.0s	+08° 53′ 32″	4.29
ν	4h 03m 09.3s	+05° 59′ 22″	3.91
ξ	3h 27m 10.1s	+09° 43′ 58″	3.74
o	3h 24m 48.7s	+09° 01′ 44″	3.60
π	4h 26m 36.4s	+14° 42′ 49″	4.69
ρ	4h 33m 50.8s	+14° 50′ 40″	4.65
σ_1	4h 39m 09.2s	+15° 47′ 59″	5.07
σ_2	4h 39m 16.4s	+15° 55′ 05″	4.69
τ	4h 42m 14.6s	+22° 57′ 25″	4.28
υ	4h 26m 18.4s	+22° 48′ 49″	4.28
φ	4h 20m 21.2s	+27° 21′ 03″	4.95
χ	4h 22m 34.9s	+25° 37′ 45″	5.37
ψ	4h 07m 00.4s	+29° 00′ 05″	5.23
ω_1	4h 09m 09.9s	+19° 36′ 33″	5.50
ω_2	4h 17m 15.6s	+20° 34′ 43″	4.94

Deep sky object	Description	Right ascension	Declination	Mag.
M1	Crab Nebula	05h 34m 31.9s	+22° 00.9′	9.0
M45	Pleiades Cluster	03h 47m 00.0s	+24° 07.0′	1.6
NGC 1647	Open Cluster	04h 45m 55.6s	+19° 06.7′	6.4
NGC 1746	Open Cluster	05h 03m 50.2s	+23° 46.1′	6.1
NGC 1807	Open Cluster	05h 10m 41.1s	+16° 31.9′	7.0
NGC 1817	Open Cluster	05h 12m 26.3s	+16° 41.0′	7.7

Taurus

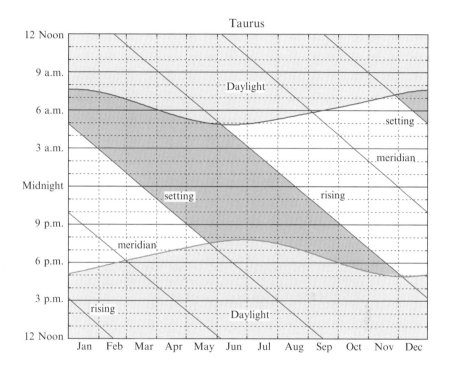

Telescopium
the Telescope

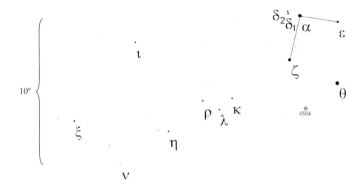

Star	Right ascension	Declination	Mag.
α	18h 26m 58.3s	−45° 58′ 06″	3.51
δ₁	18h 31m 45.3s	−45° 54′ 54″	4.96
δ₂	18h 32m 01.9s	−45° 45′ 26″	5.07
ε	18h 11m 13.7s	−45° 57′ 15″	4.53
ζ	18h 28m 49.8s	−49° 04′ 15″	4.13
η	19h 22m 51.0s	−54° 25′ 25″	5.05
ι	19h 35m 12.8s	−48° 05′ 57″	4.90
κ	18h 52m 39.5s	−52° 06′ 27″	5.17
λ	18h 58m 27.6s	−52° 56′ 18″	4.87
μ	19h 30m 34.4s	−55° 06′ 36″	6.30
ν	19h 48m 01.1s	−56° 21′ 46″	5.35
ξ	20h 07m 23.1s	−52° 52′ 51″	4.94
ρ	19h 06m 19.8s	−52° 20′ 27″	5.16

Deep sky object	Description	Right ascension	Declination	Mag.
NGC 6584	Globular Cluster	18h 18m 36.0s	−52° 13.0′	9.2

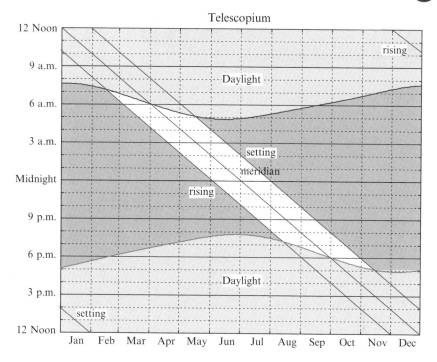

Telescopium

Triangulum

the Triangle

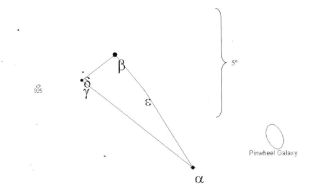

Pinwheel Galaxy

Star	Right ascension	Declination	Mag.
α	1h 53m 04.8s	+29° 34′ 44″	3.41
β	2h 09m 32.5s	+34° 59′ 14″	3.00
γ	2h 17m 18.8s	+33° 50′ 50″	4.01
δ	2h 17m 03.2s	+34° 13′ 28″	4.87
ε	2h 02m 57.9s	+33° 17′ 03″	5.50

Deep sky object	Description	Right ascension	Declination	Mag.
M33	The Pinwheel Galaxy	01h 33m 50.8s	+30° 39.6′	5.7
NGC 925	Galaxy	02h 27m 17.1s	+33° 34.7′	10.0

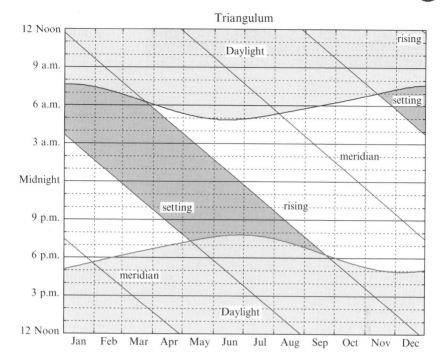

Triangulum Australe
the Southern Triangle

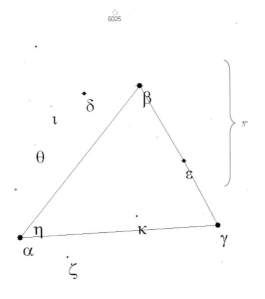

Star	Right ascension	Declination	Mag.
α	16h 48m 39.9s	−69° 01′ 40″	1.92
β	15h 55m 08.4s	−63° 25′ 50″	2.85
γ	15h 18m 54.6s	−68° 40′ 46″	2.89
δ	16h 15m 26.2s	−63° 41′ 08″	3.85
ε	15h 36m 43.1s	−66° 19′ 02″	4.11
ζ	16h 28m 28.1s	−70° 05′ 04″	4.91
η	16h 41m 23.3s	−68° 17′ 46″	5.91
θ	16h 35m 44.7s	−65° 29′ 44″	5.52
ι	16h 27m 57.2s	−64° 03′ 29″	5.27
κ	15h 55m 29.5s	−68° 36′ 11″	5.09

Deep sky object	Description	Right ascension	Declination	Mag.
NGC 6025	Open Cluster	16h 03m 17.0s	−60° 25.9′	5.1

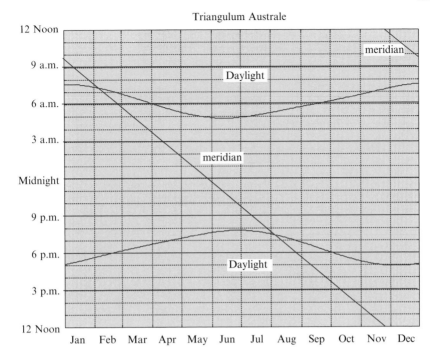

The constellation Triangulum Australe is not visible from mid-northern latitudes

Tucana

the Toucan

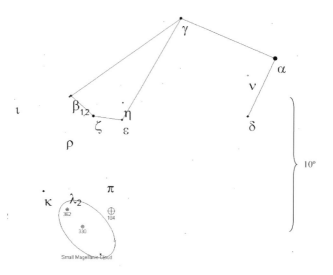

Star	Right ascension	Declination	Mag.
α	22h 18m 30.1s	−60° 15′ 35″	2.86
β₁	00h 31m 32.7s	−62° 57′ 30″	4.37
β₂	00h 31m 33.6s	−62° 57′ 57″	4.54
β₃	00h 32m 43.8s	−63° 01′ 53″	5.09
γ	23h 17m 25.7s	−58° 14′ 08″	3.99
δ	22h 27m 20.0s	−64° 58′ 00″	4.48
ε	23h 59m 55.0s	−65° 34′ 38″	4.50
ζ	00h 20m 04.2s	−64° 52′ 30″	4.23
η	23h 57m 35.2s	−64° 17′ 55″	5.00
θ	00h 33m 23.4s	−71° 16′ 00″	6.13
ι	01h 07m 18.7s	−61° 46′ 31″	5.37
κ	01h 15m 46.2s	−68° 52′ 34″	4.86
λ₁	00h 52m 24.4s	−69° 30′ 16″	6.22
λ₂	00h 55m 00.3s	−69° 31′ 38″	5.45
ν	22h 33m 00.0s	−61° 58′ 57″	4.81
π	00h 20m 38.9s	−69° 37′ 30″	5.51
ρ	00h 42m 28.4s	−65° 28′ 05″	5.39

Deep sky object	Description	Right ascension	Declination	Mag.
NGC 104	Globular Cluster	00h 24m 05.1s	−72° 04.8′	4.0
NGC 292	Small Magellanic Cloud	00h 52m 38.0s	−72° 48.0′	2.3
NGC 330	Globular Cluster	00h 56m 19.8s	−72° 27.7′	9.6
NGC 362	Globular Cluster	01h 03m 14.2s	−70° 50.9′	6.6

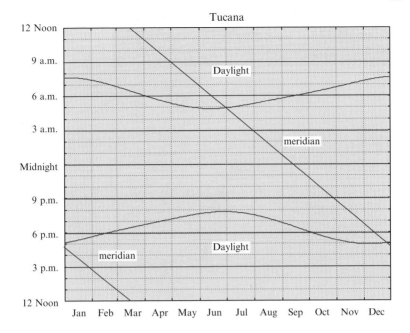

The constellation Tucana is not visible from mid-northern latitudes

Ursa Major
the Great Bear

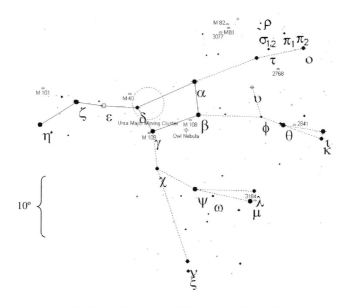

Star	Right ascension	Declination	Mag.
α	11h 03m 43.6s	+61° 45′ 03″	1.79
β	11h 01m 50.4s	+56° 22′ 56″	2.37
γ	11h 53m 49.8s	+53° 41′ 41″	2.44
δ	12h 15m 25.5s	+57° 01′ 57″	3.31
ε	12h 54m 01.7s	+55° 57′ 35″	1.77
ζ₁	13h 23m 55.5s	+54° 55′ 31″	2.27
ζ₂	13h 23m 56.3s	+54° 55′ 18″	3.95
η	13h 47m 32.3s	+49° 18′ 48″	1.86
θ	09h 32m 51.3s	+51° 40′ 38″	3.17
ι	08h 59m 12.4s	+48° 02′ 30″	3.14
κ	09h 03m 37.5s	+47° 09′ 24″	3.60
λ	10h 17m 05.7s	+42° 54′ 52″	3.45
μ	10h 22m 19.7s	+41° 29′ 58″	3.05
ν	11h 18m 28.7s	+33° 05′ 39″	3.48
ξ₁	11h 18m 10.9s	+31° 31′ 45″	4.87
ξ₂	11h 18m 10.9s	+31° 31′ 45″	4.41
o	08h 30m 15.8s	+60° 43′ 05″	3.36
π₁	08h 39m 11.7s	+65° 01′ 15″	5.64
π₂	08h 40m 12.9s	+64° 19′ 41″	4.60
ρ	09h 02m 32.7s	+67° 37′ 47″	4.76
σ₁	09h 08m 23.5s	+66° 52′ 24″	5.14
σ₂	09h 10m 23.1s	+67° 08′ 04″	4.80
τ	09h 10m 55.0s	+63° 30′ 49″	4.67
υ	09h 50m 59.3s	+59° 02′ 19″	3.80
φ	09h 52m 06.3s	+54° 03′ 51″	4.59
χ	11h 46m 03.0s	+47° 46′ 46″	3.71
ψ	11h 09m 39.7s	+44° 29′ 54″	3.01
ω	10h 53m 58.7s	+43° 11′ 24″	4.71

Deep sky object	Description	Right ascension	Declination	Mag.
M40	Double Star	12h 22m 24.0s	+58° 05.0'	8.4
M81	Bode's Galaxy	09h 55m 32.9s	+69° 03.9'	6.8
M82	The Cigar Galaxy	09h 55m 50.7s	+69° 40.7'	8.4
M97	The Owl Nebula	11h 14m 47.7s	+55° 01.0'	7.7
M101	The Pinwheel Galaxy	14h 03m 12.0s	+54° 21.0'	7.7
M108	Galaxy	11h 11m 31.2s	+55° 40.4'	10.0
M109	Galaxy	11h 57m 36.0s	+53° 23.5'	9.8
NGC 2768	Galaxy	09h 11m 37.5s	+60° 02.2'	10.0
NGC 2841	Spiral Galaxy	09h 22m 02.3s	+50° 58.7'	9.3
NGC 3077	Irregular Galaxy	10h 03m 20.4s	+68° 44.0'	9.9
NGC 3184	Spiral Galaxy	10h 18m 17.0s	+41° 25.5'	9.8

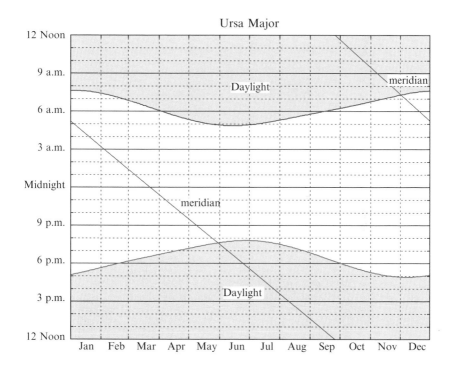

Ursa Major

Ursa Minor
the Smaller Bear

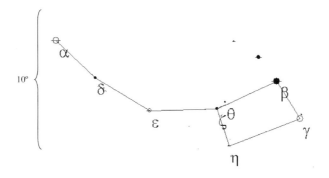

Star	Right ascension	Declination	Mag.
α	02h 31m 50.5s	+89° 15′ 51″	2.02
β	14h 50m 42.2s	+74° 09′ 20″	2.08
γ	15h 20m 43.6s	+71° 50′ 02″	3.05
δ	17h 32m 12.5s	+86° 35′ 11″	4.36
ε	16h 45m 57.8s	+82° 02′ 14″	4.23
ζ	15h 44m 03.3s	+77° 47′ 40″	4.32
η	16h 17m 30.2s	+75° 45′ 19″	4.95
θ	15h 31m 24.7s	+77° 20′ 57″	4.96
λ	17h 16m 56.0s	+89° 02′ 15″	6.38

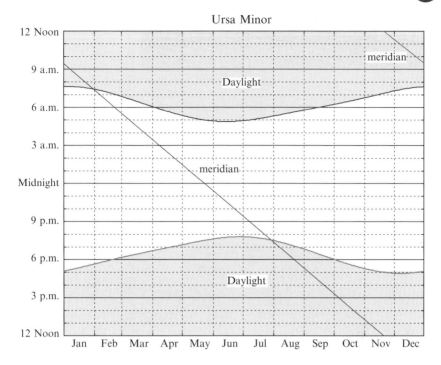

Vela

the Ship's Sail

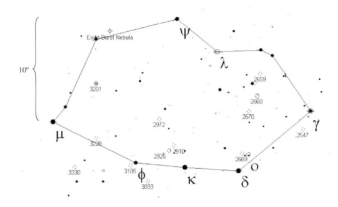

Star	Right ascension	Declination	Mag.
γ_1	08h 09m 29.2s	−47° 20′ 44″	4.27
γ_2	08h 09m 31.9s	−47° 20′ 12″	1.78
δ	08h 44m 42.2s	−54° 42′ 30″	1.96
κ	09h 22m 06.8s	−55° 00′ 38″	2.50
λ	09h 07m 59.7s	−43° 25′ 57″	2.21
μ	10h 46m 46.1s	−49° 25′ 12″	2.69
o	08h 40m 17.6s	−52° 55′ 19″	3.62
φ	09h 56m 51.7s	−54° 34′ 04″	3.54
ψ	09h 30m 41.9s	−40° 28′ 00″	3.60

Deep sky object	Description	Right ascension	Declination	Mag.
NGC 2547	Open Cluster	08h 10m 10.5s	−49° 13.5′	4.7
NGC 2659	Open Cluster	08h 42m 33.0s	−45° 00.0′	8.6
NGC 2660	Open Cluster	08h 42m 38.0s	−47° 12.0′	8.8
NGC 2669	Open Cluster	08h 46m 22.0s	−52° 56.9′	6.1
NGC 2670	Open Cluster	08h 45m 29.5s	−48° 47.5′	7.8
NGC 2910	Open Cluster	09h 30m 29.0s	−52° 54.8′	7.2
NGC 2925	Open Cluster	09h 33m 10.9s	−53° 23.8′	8.3
NGC 2972	Open Cluster	09h 40m 11.5s	−50° 19.3′	9.9
NGC 3033	Open Cluster	09h 48m 39.1s	−56° 24.7′	8.8
NGC 3105	Open Cluster	10h 00m 39.5s	−54° 47.3′	9.7
NGC 3132	Eight Burst Nebula	10h 07m 01.8s	−40° 26.2′	8.0
NGC 3201	Globular Cluster	10h 17m 36.7s	−46° 24.7′	6.7
NGC 3228	Open Cluster	10h 21m 22.2s	−51° 43.9′	6.0
NGC 3330	Open Cluster	10h 38m 47.5s	−54° 06.9′	7.4

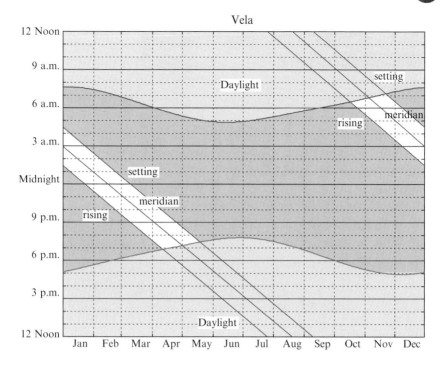
Vela

Virgo

the Maiden

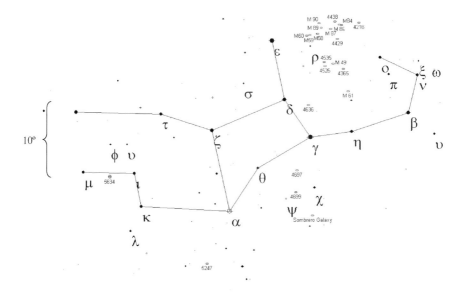

Star	Right ascension	Declination	Mag.
α	13h 25m 11.5s	−11° 09′ 41″	0.98
β	11h 50m 41.6s	+01° 45′ 53″	3.61
γ₁	12h 41m 39.5s	−01° 26′ 58″	3.65
γ₂	12h 41m 39.5s	−01° 26′ 58″	3.68
δ	12h 55m 36.1s	+03° 23′ 51″	3.38
ε	13h 02m 10.5s	+10° 57′ 33″	2.83
ζ	13h 34m 41.5s	−00° 35′ 46″	3.37
η	12h 19m 54.3s	−00° 40′ 00″	3.89
θ	13h 09m 56.9s	−05° 32′ 20″	4.38
ι	14h 16m 00.8s	−06° 00′ 02″	4.08
κ	14h 12m 53.7s	−10° 16′ 25″	4.19
λ	14h 19m 06.5s	−13° 22′ 16″	4.52
μ	14h 43m 03.5s	−05° 39′ 30″	3.88
ν	11h 45m 51.5s	+06° 31′ 46″	4.03
ξ	11h 45m 17.0s	+08° 15′ 30″	4.85
o	12h 05m 12.5s	+08° 43′ 59″	4.12
π	12h 00m 52.3s	+06° 36′ 51″	4.66
ρ	12h 41m 53.0s	+10° 14′ 08″	4.88
σ	13h 17m 36.2s	+05° 28′ 12″	4.80
τ	14h 01m 38.7s	+01° 32′ 40″	4.26
υ	14h 19m 32.4s	−02° 15′ 56″	5.14
φ	14h 28m 12.1s	−02° 13′ 41″	4.81
χ	12h 39m 14.7s	−07° 59′ 44″	4.66
ψ	12h 54m 21.1s	−09° 32′ 20″	4.79
ω	11h 38m 27.5s	+08° 08′ 03″	5.36

Deep sky object	Description	Right ascension	Declination	Mag.
M49	Galaxy	12h 29m 46.7s	+08° 00.0′	8.4
M58	Galaxy	12h 37m 43.5s	+11° 49.1′	9.8
M59	Galaxy	12h 42m 00.0s	+11° 39.0′	9.8
M60	Galaxy	12h 43m 42.0s	+11° 33.0′	8.8
M61	Galaxy	12h 21m 54.9s	+04° 28.4′	9.7
M84	Galaxy	12h 25m 04.7s	+12° 53.2′	9.1
M86	Galaxy	12h 26m 11.9s	+12° 56.8′	8.9
M87	Galaxy	12h 30m 49.3s	+12° 23.4′	8.6
M89	Galaxy	12h 35m 39.9s	+12° 33.4′	9.8
M90	Galaxy	12h 36m 49.9s	+13° 09.8′	9.5
M104	Sombrero Galaxy	12h 39m 59.4s	−11° 37.4′	8.3
NGC 4216	Edge-on Spiral Galaxy	12h 15m 54.1s	+13° 09.0′	10.0
NGC 4365	Galaxy	12h 24m 28.1s	+07° 19.1′	9.6
NGC 4438	Galaxy	12h 27m 45.6s	+13° 00.5′	10.0
NGC 4526	Galaxy	12h 34m 03.1s	+07° 42.0′	9.6
NGC 4535	Spiral Galaxy	12h 34m 20.3s	+08° 11.9′	10.0
NGC 4636	Galaxy	12h 42m 48.0s	+02° 41.0′	9.6
NGC 4697	Galaxy	12h 48m 36.0s	−05° 48.0′	9.3
NGC 4699	Galaxy	12h 49m 00.0s	−08° 40.0′	9.6
NGC 5247	Galaxy	13h 38m 06.0s	−17° 53.0′	10.0
NGC 5634	Globular Cluster	14h 29m 36.0s	−05° 59.0′	9.4

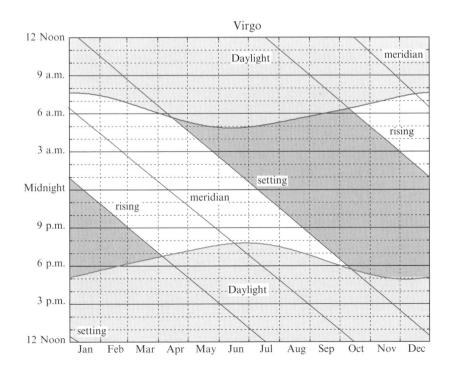

Volans

the Flying Fish

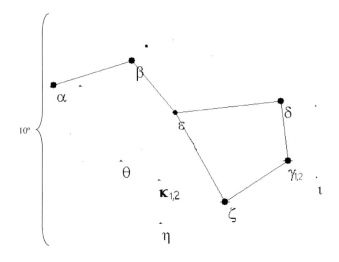

Star	Right ascension	Declination	Mag.
α	09h 02m 26.9s	−66° 23' 46"	4.00
β	08h 25m 44.3s	−66° 08' 13"	3.77
γ_1	07h 08m 42.2s	−70° 29' 50"	5.67
γ_2	07h 08m 45.0s	−70° 29' 57"	3.78
δ	07h 16m 49.8s	−67° 57' 27"	3.98
ε	08h 07m 55.9s	−68° 37' 02"	4.35
ζ	07h 41m 49.3s	−72° 36' 22"	3.95
η	08h 22m 04.7s	−73° 24' 01"	5.29
θ	08h 39m 05.3s	−70° 23' 13"	5.20
ι	06h 51m 27.0s	−70° 57' 49"	5.40
κ_1	08h 19m 49.2s	−71° 30' 54"	5.37
κ_2	08h 20m 00.8s	−71° 30' 19"	5.65

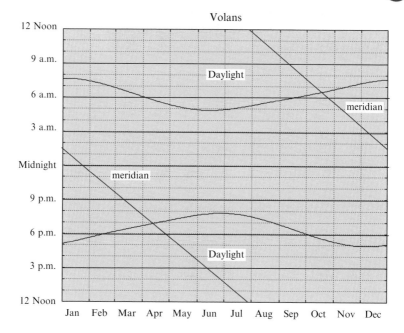

The constellation Volans is not visible from mid-northern latitudes

Vulpecula

the Fox

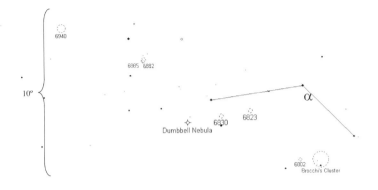

Star	Right ascension	Declination	Mag.
α	19h 28m 42.2s	+24° 39′ 54″	4.44

Deep sky object	Description	Right ascension	Declination	Mag.
M27	The Dumbbell Nebula	19h 59m 36.5s	+22° 43.2′	7.3
NGC 6802	Open Cluster	19h 30m 36.0s	+20° 16.0′	8.8
NGC 6823	Open Cluster	19h 43m 06.0s	+23° 18.0′	7.1
NGC 6830	Open Cluster	19h 51m 00.0s	+23° 04.0′	7.9
NGC 6882	Open Cluster	20h 11m 42.0s	+26° 33.0′	8.1
NGC 6885	Open Cluster	20h 12m 00.0s	+26° 29.0′	8.1
NGC 6940	Open Cluster	20h 34m 36.0s	+28° 18.0′	6.3
Collinder 399	Brocchi's Cluster (Coathanger)	19h 25m 24.0s	+20° 11.0′	3.6

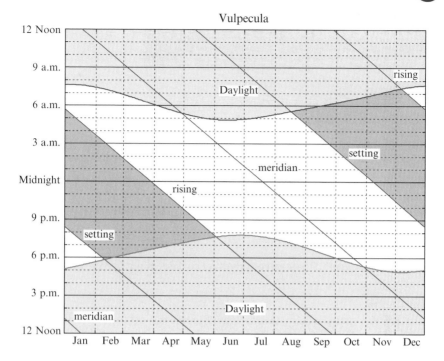

Vulpecula

Constellations near the meridian in January

7 p.m. – 9 p.m.
Aries, Caelum, Camelopardalis, Cetus, Dorado, Eridanus, Fornax, Horologium, Hydrus, Perseus, Reticulum, Taurus, Triangulum

9 p.m. – 11 p.m.
Auriga, Caelum, Camelopardalis, Canis Major, Columba, Dorado, Lepus, Mensa, Orion, Pictor, Reticulum, Taurus

11 p.m. – 1 a.m.
Cancer, Canis Major, Canis Minor, Columba, Gemini, Lynx, Monoceros, Pictor, Puppis, Pyxis, Volans

1 a.m. – 3 a.m.
Antlia, Cancer, Carina, Chamaeleon, Hydra, Leo, Leo Minor, Lynx, Pyxis, Sextans, Vela, Volans

3 a.m. – 5 a.m.
Antlia, Canes Venatici, Chamaeleon, Coma Berenices, Corvus, Crater, Crux, Hydra, Leo, Leo Minor, Musca, Sextans, Ursa Major, Virgo

Constellations near the meridian in February

7 p.m. – 9 p.m.
Auriga, Caelum, Camelopardalis, Canis Major, Columba, Dorado, Lepus, Mensa, Orion, Pictor, Reticulum, Taurus

9 p.m. – 11 p.m.
Cancer, Canis Major, Canis Minor, Columba, Gemini, Lynx, Monoceros, Pictor, Puppis, Pyxis, Volans

11 p.m. – 1 a.m.
Antlia, Cancer, Carina, Chamaeleon, Hydra, Leo, Leo Minor, Lynx, Pyxis, Sextans, Vela, Volans

1 a.m. – 3 a.m.
Antlia, Canes Venatici, Chamaeleon, Coma Berenices, Corvus, Crater, Crux, Hydra, Leo, Leo Minor, Musca, Sextans, Ursa Major, Virgo

3 a.m. – 5 a.m.
Boötes, Canes Venatici, Centaurus, Circinus, Coma Berenices, Corvus, Crux, Musca, Virgo

Constellations near the meridian in March

7 p.m. – 9 p.m.
Cancer, Canis Major, Canis Minor, Columba, Gemini, Lynx, Monoceros, Pictor, Puppis, Pyxis, Volans

9 p.m. – 11 p.m.
Antlia, Cancer, Carina, Chamaeleon, Hydra, Leo, Leo Minor, Lynx, Pyxis, Sextans, Vela, Volans

11 p.m. – 1 a.m.
Antlia, Canes Venatici, Chamaeleon, Coma Berenices, Corvus, Crater, Crux, Hydra, Leo, Leo Minor, Sextans, Musca, Ursa Major, Virgo

1 a.m. – 3 a.m.
Boötes, Canes Venatici, Centaurus, Circinus, Coma Berenices, Corvus, Crux, Musca, Virgo

3 a.m. – 5 a.m.
Apus, Boötes, Circinus, Corona Borealis, Draco, Hercules, Libra, Lupus, Norma, Scorpius, Serpens Caput, Triangulum Australe, Ursa Minor

Constellations near the meridian in April

7 p.m. – 9 p.m.
Antlia, Cancer, Carina, Chamaeleon, Hydra, Leo, Leo Minor, Lynx, Pyxis, Sextans, Vela, Volans

9 p.m. – 11 p.m.
Antlia, Canes Venatici, Chamaeleon, Coma Berenices, Corvus, Crater, Crux, Hydra, Leo, Leo Minor, Musca, Sextans, Ursa Major, Virgo

11 p.m. – 1 a.m.
Boötes, Canes Venatici, Centaurus, Circinus, Coma Berenices, Corvus, Crux, Musca, Virgo

1 a.m. – 3 a.m.
Apus, Boötes, Circinus, Corona Borealis, Draco, Hercules, Libra, Lupus, Norma, Scorpius, Serpens Caput, Triangulum Australe, Ursa Minor

3 a.m. – 5 a.m.
Ara, Hercules, Norma, Ophiuchus, Sagittarius, Scorpius, Scutum, Serpens Caput, Serpens Cauda, Telescopium, Triangulum Australe

Constellations near the meridian in May

7 p.m. – 9 p.m.
Antlia, Canes Venatici, Chamaeleon, Coma Berenices, Corvus, Crater, Crux, Hydra, Leo, Leo Minor, Musca, Sextans, Ursa Major, Virgo

9 p.m. – 11 p.m.
Boötes, Canes Venatici, Centaurus, Circinus, Coma Berenices, Corvus, Crux, Musca, Virgo

11 p.m. – 1 a.m.
Apus, Boötes, Corona Borealis, Circinus, Draco, Hercules, Libra, Lupus, Norma, Scorpius, Serpens Caput, Triangulum Australe, Ursa Minor

1 a.m. – 3 a.m.
Ara, Hercules, Norma, Ophiuchus, Sagittarius, Scorpius, Scutum, Serpens Caput, Serpens Cauda, Telescopium, Triangulum Australe

3 a.m. – 5 a.m.
Aquila, Cepheus, Corona Australis, Cygnus, Delphinus, Lyra, Octans, Pavo, Sagitta, Sagittarius, Scutum, Serpens Cauda, Telescopium, Vulpecula

Constellations near the meridian in June

7 p.m. – 9 p.m.
Boötes, Canes Venatici, Centaurus, Circinus, Coma Berenices, Corvus, Crux, Musca, Virgo

9 p.m. – 11 p.m.
Apus, Boötes, Circinus, Corona Borealis, Draco, Hercules, Libra, Lupus, Norma, Scorpius, Serpens Caput, Triangulum Australe, Ursa Minor

11 p.m. – 1 a.m.
Ara, Hercules, Norma, Ophiuchus, Sagittarius, Scorpius, Scutum, Serpens Caput, Serpens Cauda, Telescopium, Triangulum Australe

1 a.m. – 3 a.m.
Aquila, Cepheus, Corona Australis, Cygnus, Delphinus, Lyra, Octans, Pavo, Sagitta, Sagittarius, Scutum, Serpens Cauda, Telescopium, Vulpecula

3 a.m. – 5 a.m.
Aquarius, Capricornus, Cepheus, Cygnus, Delphinus, Equuleus, Grus, Indus, Lacerta, Microscopium, Piscis Austrinus, Sagitta

Constellations near the meridian in July

7 p.m. – 9 p.m.
Apus, Boötes, Circinus, Corona Borealis, Draco, Hercules, Libra, Lupus, Norma, Scorpius, Serpens Caput, Triangulum Australe, Ursa Minor

9 p.m. – 11 p.m.
Ara, Hercules, Norma, Ophiuchus, Sagittarius, Scorpius, Scutum, Serpens Caput, Serpens Cauda, Telescopium, Triangulum Australe

11 p.m. – 1 a.m.
Aquila, Cepheus, Corona Australis, Cygnus, Delphinus, Lyra, Octans, Pavo, Sagitta, Sagittarius, Scutum, Serpens Cauda, Telescopium, Vulpecula

1 a.m. – 3 a.m.
Aquarius, Capricornus, Cepheus, Cygnus, Delphinus, Equuleus, Grus, Indus, Lacerta, Microscopium, Piscis Austrinus, Sagitta

3 a.m. – 5 a.m.
Andromeda, Aquarius, Cassiopeia, Grus, Lacerta, Pegasus, Phoenix, Pisces, Piscis Austrinus, Sculptor, Tucana

Constellations near the meridian in August

7 p.m. – 9 p.m.
Ara, Hercules, Norma, Ophiuchus, Sagittarius, Scorpius, Scutum, Serpens Caput, Serpens Cauda, Telescopium, Triangulum Australe

9 p.m. – 11 p.m.
Aquila, Cepheus, Corona Australis, Cygnus, Delphinus, Lyra, Octans, Pavo, Sagitta, Sagittarius, Scutum, Serpens Cauda, Telescopium, Vulpecula

11 p.m. – 1 a.m.
Aquarius, Capricornus, Cepheus, Cygnus, Delphinus, Equuleus, Grus, Indus, Lacerta, Microscopium, Piscis Austrinus, Sagitta

1 a.m. – 3 a.m.
Andromeda, Aquarius, Cassiopeia, Grus, Lacerta, Pegasus, Phoenix, Pisces, Piscis Austrinus, Sculptor, Tucana

3 a.m. – 5 a.m.
Andromeda, Aries, Cassiopeia, Cetus, Fornax, Hydrus, Phoenix, Sculptor, Triangulum

Constellations near the meridian in September

7 p.m. – 9 p.m.
Aquila, Cepheus, Corona Australis, Cygnus, Delphinus, Lyra, Octans, Pavo, Sagitta, Sagittarius, Scutum, Serpens Cauda, Telescopium, Vulpecula

9 p.m. – 11 p.m.
Aquarius, Capricornus, Cepheus, Cygnus, Delphinus, Equuleus, Grus, Indus, Lacerta, Microscopium, Piscis Austrinus, Sagitta

11 p.m. – 1 a.m.
Andromeda, Aquarius, Cassiopeia, Grus, Lacerta, Pegasus, Phoenix, Pisces, Piscis Austrinus, Sculptor, Tucana

1 a.m. – 3 a.m.
Andromeda, Aries, Cassiopeia, Cetus, Fornax, Hydrus, Phoenix, Sculptor, Triangulum

3 a.m. – 5 a.m.
Aries, Caelum, Camelopardalis, Cetus, Dorado, Eridanus, Fornax, Horologium, Perseus, Hydrus, Reticulum, Taurus, Triangulum

Constellations near the meridian in October

7 p.m. – 9 p.m.
Aquarius, Capricornus, Cepheus, Cygnus, Delphinus, Equuleus, Grus, Indus, Lacerta, Microscopium, Piscis Austrinus, Sagitta

9 p.m. – 11 p.m.
Andromeda, Aquarius, Cassiopeia, Grus, Lacerta, Pegasus, Phoenix, Pisces, Piscis Austrinus, Sculptor, Tucana

11 p.m. – 1 a.m.
Andromeda, Aries, Cassiopeia, Cetus, Fornax, Hydrus, Phoenix, Sculptor, Triangulum

1 a.m. – 3 a.m.
Aries, Caelum, Camelopardalis, Cetus, Dorado, Eridanus, Fornax, Horologium, Hydrus, Perseus, Reticulum, Taurus, Triangulum

3 a.m. – 5 a.m.
Auriga, Caelum, Camelopardalis, Canis Major, Columba, Dorado, Lepus, Mensa, Orion, Pictor, Reticulum, Taurus

Constellations near the meridian in November

7 p.m. – 9 p.m.
Andromeda, Aquarius, Cassiopeia, Grus, Lacerta, Pegasus, Phoenix, Pisces, Piscis Austrinus, Sculptor, Tucana

9 p.m. – 11 p.m.
Andromeda, Aries, Cassiopeia, Cetus, Fornax, Hydrus, Phoenix, Sculptor, Triangulum

11 p.m. – 1 a.m.
Aries, Caelum, Camelopardalis, Cetus, Dorado, Eridanus, Fornax, Horologium, Hydrus, Perseus, Reticulum, Taurus, Triangulum

1 a.m. – 3 a.m.
Auriga, Caelum, Camelopardalis, Canis Major, Columba, Dorado, Lepus, Mensa, Orion, Pictor, Reticulum, Taurus

3 a.m. – 5 a.m.
Cancer, Canis Major, Canis Minor, Columba, Gemini, Lynx, Monoceros, Pictor, Puppis, Pyxis, Volans

Constellations near the meridian in December

7 p.m. – 9 p.m.
Andromeda, Aries, Cassiopeia, Cetus, Fornax, Hydrus, Phoenix, Sculptor, Triangulum

9 p.m. – 11 p.m.
Aries, Caelum, Camelopardalis, Cetus, Dorado, Eridanus, Fornax, Horologium, Hydrus, Perseus, Reticulum, Taurus, Triangulum

11 p.m. – 1 a.m.
Auriga, Caelum, Camelopardalis, Canis Major, Columba, Dorado, Lepus, Mensa, Orion, Pictor, Reticulum, Taurus

1 a.m. – 3 a.m.
Cancer, Canis Major, Canis Minor, Columba, Gemini, Lynx, Monoceros, Pictor, Puppis, Pyxis, Volans

3 a.m. – 5 a.m.
Antlia, Cancer, Carina, Chamaeleon, Hydra, Leo, Leo Minor, Lynx, Pyxis, Sextans, Vela, Volans

About the Author

Jeff A. Farinacci became interested in astronomy at the age of 4 when his brother Jim showed him Comet Ikeya-Seki. He graduated from the University of Utah with two Bachelor of Science degrees (both Magna Cum Laude), in Mathematics and Physics. He also graduated from Indiana University with a Master of Arts degree in Mathematics. He has worked as a Software Engineer on a variety of projects for various companies. He has worked on a variety of defense projects, wrote various simulations, and wrote a 3D graphics library for a graphics chip and designed the math functions that went onto another chip.

He also wrote "Stellar Guides for Your Birthday," an article in the November 2001 issue of Sky & Telescope magazine.

Other Titles in This Series *(continued from page ii)*

Printed in the United States of America